Optimal Crossover Designs

Optimal Crossover Designs

Mausumi Bose
Indian Statistical Institute, Kolkata, India

Aloke Dey
Indian Statistical Institute, New Delhi, India

World Scientific

NEW JERSEY • LONDON • SINGAPORE • BEIJING • SHANGHAI • HONG KONG • TAIPEI • CHENNAI

Published by

World Scientific Publishing Co. Pte. Ltd.

5 Toh Tuck Link, Singapore 596224

USA office: 27 Warren Street, Suite 401-402, Hackensack, NJ 07601

UK office: 57 Shelton Street, Covent Garden, London WC2H 9HE

Library of Congress Cataloging-in-Publication Data

Bose, Mausumi.

Optimal crossover designs / by Mausumi Bose & Aloke Dey.

p. cm.

Includes bibliographical references and index.

ISBN-13: 978-981-281-842-3 (hardcover : alk. paper)

ISBN-10: 981-281-842-1 (hardcover : alk. paper)

1. Experimental design. 2. Optimal designs (Statistics) I. Dey, Aloke. II. Title.

QA279.B666 2009

519.5'7--dc22

2009007076

British Library Cataloguing-in-Publication Data

A catalogue record for this book is available from the British Library.

Printed in Singapore.

To my parents, MB

To my family, AD

Contents

Preface

Among a variety of experimental designs that are available for treatment comparison experiments, the crossover designs, also known as change-over or repeated measurements designs, occupy an important place. These designs are widely used in many diverse areas such as animal nutrition experiments, pharmaceutical studies and clinical trials, biological assays, weather modification experiments, sensory evaluation of food products, psychology, bio-equivalence studies, consumer trials, questionnaire-based surveys, and so on. In particular, over the last few years, the use of crossover designs in pharmaceutical studies and clinical trials has been quite extensive. Although there is a long history of the use of these designs in experiments, the study of their optimality aspects began only about thirty years ago. During the last three decades, the area of optimal crossover designs has seen a vigorous growth. The literature is already voluminous and is still growing.

This rich body of research findings on optimal crossover designs is scattered across various issues of journals and there is no single book currently available that deals exclusively with this topic. We hope that this book will fill this gap by presenting a comprehensive and up-to-date account of the major advances in optimal crossover designs. We have attempted to cover all those developments in this area which, in our perception, are the most significant ones. However, this book is not intended as an encyclopedic compendium on crossover designs as our focus is only on the optimality aspects of such designs.

The book is organized into seven chapters. Chapter 1 is introductory in nature and describes the notation and terminology to be followed throughout the book. A brief description of the contents of the remaining chapters also appears in Chapter 1. The traditional model, i.e., the homoscedastic fixed effects crossover model with first order carryover effects, is introduced in Chapter 1. In Chapter 2, we review optimality results under this traditional model for designs where the number of periods may equal or exceed the number of treatments under study. Chapter 3 also comprises optimality results under this same model, but in contrast to Chapter 2, here we consider the situation where the number of periods can be at most equal to the number of treatments. In Chapter 4, results on optimal crossover designs via an approximate design theory are presented. Chapters 2–4 essentially consider the traditional model while in Chapters 5 and 6, more complex models for crossover designs are considered and optimality results under these models are described. Finally, in Chapter 7, some further developments in this area are discussed. Most of the results in the chapters are illustrated through a large number of examples.

A background in basic matrix algebra and linear statistical models is assumed. We also assume that the reader is familiar with the general area of design of experiments at a graduate level.

We sincerely thank Rahul Mukerjee for going through the manuscript and making numerous suggestions for improvement. The work of A. Dey was supported by the Indian National Science Academy under the Senior Scientist Scheme of the Academy. This support is gratefully acknowledged. We also thank Ms. E. H. Chionh, Ms. Jessie Tan and the editorial and production staff of World Scientific Publishing for their prompt and excellent handling of the project. Our special thanks to our families for their love and support. We are also thankful to the Indian Statistical Institute for providing a conducive environment for carrying out this work.

Mausumi Bose
Aloke Dey

Kolkata
New Delhi
January 2009

Chapter 1

Introduction

1.1 Prologue

The subject of design of experiments has seen a phenomenal growth since the time R. A. Fisher laid the modern foundations of this important discipline during the early part of the 20th century. It has played, and continues to play, a major role in almost all areas of scientific investigation including agriculture and animal husbandry, biology, medicine, physical and chemical sciences and industrial research. Design of experiments is an area of intensive research and forms an integral part of most statistics curricula.

Among many designs that are available for treatment comparison experiments, the *crossover* designs have occupied an important place. In recent years, the application of these designs in different areas has been quite widespread and simultaneously, many important theoretical results have also been obtained. This book is mainly concerned with the optimality aspects of crossover designs, an area of much research activity in the past three decades.

Crossover designs, also known as *change-over* or *repeated measurements* designs, have long been used in several diverse areas. Apparently, the use of crossover trials was first made in agriculture in the 19th century. For a brief history of the early use of crossover trials in agriculture and other areas, one may refer to Jones and Kenward (2003, Section 1.4). Applications of crossover designs in agriculture and animal feeding trials were made in the early part of the 20th century by Cochran (1939) and Cochran, Autrey and Cannon (1941). Cochran (1939) considered the problems associated with the planning of long-term crop rotation experiments and was the first to separate out the direct and carryover effects. Cochran *et al.* (1941) used a type of crossover design for feeding trials in dairy cattle. Another early paper on the application and analysis of crossover trials in animal feeding experiments is due to Patterson (1951).

Crossover designs have now found applications in several other branches of science like pharmaceutical studies and clinical trials (Senn (2003)), biological assays (Finney (1956, 1978)), weather modification experiments (Mielke (1974)), sensory evaluation of food products (Durier, Monod and Bruetschy (1997), Kunert (1998)), psychology (Namboodiri (1972), Cotton (1998)), bio-equivalence studies (Chow and Liu (1992)), consumer trials (Wakeling and MacFie (1995)) and questionnaire-based surveys (Lakatos and Raghavarao (1987)). In particular, over the last few years, the use of crossover designs in pharmaceutical studies and clinical trials has been quite extensive. Real life examples and discussion can be found in books on design of experiments, e.g., by Federer (1955), Cochran and Cox (1957), Cox (1958), John and Quenouille (1977) and Hinkelmann and Kempthorne (2005). In recent years, several books dealing exclusively with crossover trials have appeared, e.g., by Ratkowsky, Evans and Alldredge (1992), Jones and Kenward (2003) and Senn (2003), where more detailed descriptions of the design, analysis and practical applications can be found. Brief descriptions of the design and analysis of crossover trials can be found in encyclopedia articles by Kenward and Jones (1998) and Senn (1997, 2000). Excellent reviews on the subject from different standpoints, including discussions on practical applications and additional references, have appeared at different points of time and, for such reviews, we refer to Hedayat and Afsarinejad (1975), Hedayat (1981), Bishop and Jones (1984), Matthews (1988), Afsarinejad (1990b), Stufken (1996), Jones and Deppe (2001) and Bate and Jones (2008). The authoritative review of Stufken (1996) is on optimal crossover designs and is particularly relevant for this book.

Unlike other conventional designs, in crossover designs one uses an experimental unit for measurements over different occasions. The *units*, often called *subjects*, could be humans, animals, plots of land, etc. Henceforth, we shall generally use the term "subjects", rather than "units". The different occasions at which the subjects are used are referred to as *periods*. Thus, a typical crossover trial (or experiment) involves n experimental subjects, each subject being observed for p periods, resulting in a total of np experimental sites. Suppose such a trial aims at drawing inference on t treatments. Any allocation of these t treatments to the np experimental sites is called a crossover design. A crossover design as described above thus yields np observations, one measurement being made on each experimental site. Note that even though there are only n experimental subjects in a crossover trial, the number of observations generated is np since each subject gives one observation from each of the p periods. Throughout this book, a crossover design with n subjects and p periods will be displayed as a $p \times n$ array, with rows of the array representing the periods and columns representing the subjects. Also, the treatments will be denoted by the numbers 1,2, ... or by the letters A, B, ... The following are some illustrations indicating the application of crossover designs in diverse fields. These are merely illustrative and, by no means exhaustive.

Example 1.1.1. The simplest and widely used crossover design is the two-treatment, two-period design, commonly known as the AB/BA design. Here A and B are the two treatments; for example, in the context of clinical trials, these could be two drugs or, a drug and a placebo. Furthermore, over the two periods, any subject is assigned one of the two distinct treatment sequences, namely, A followed by B, that is, AB, or B followed by A, that is, BA. For a review of the AB/BA trials in clinical research and the associated statistical issues involved, see, e.g., Chassan (1964), Senn (1994) and Jones and Kenward (2003). We cite only two instances of their use.

An application of AB/BA designs in pharmacokinetic studies was described by Jones, Wang, Jarvis and Byrom (1999). Pharmacokinetics is concerned with the study of absorption, distribution, metabolism and elimination of drugs in human beings. The study considered by Jones *et al.* (1999) related to a drug for the treatment of hypertension. Two different doses of the drug were the treatments A and B, and four patients were exposed to each of the two treatment sequences, AB and BA. In another context, Rupp *et al.* (2008) reported the results of a crossover experiment conducted over a period of two years to examine the effect of an alcohol-based hand gel (A) vis-á-vis a placebo (B) in intensive care units of certain hospitals. □

Example 1.1.2. The use of crossover designs involving more than two treatments and more than two periods is also quite common. Senn and Hildebrand (1991) gave an example of a crossover experiment involving two drugs, namely, Formoterol (A), Salbutamol (B) and a placebo (C), for studying the effects of these on exercise induced asthma. There were 30 patients (subjects) available for the study and a crossover design with three periods and 6 distinct treatment sequences was used. These sequences are shown below, each column representing a treatment sequence.

A	B	C	A	B	C
B	C	A	C	A	B .
C	A	B	B	C	A

In the actual experiment, 5 patients were given the first treatment sequence, 3 patients the second sequence, 6 patients each the third and fourth sequences and 5 patients each were given the fifth and sixth sequences.

Another example of an application of a crossover design in clinical trials involving more than two treatments was given by Willey, Grant and Pocock (1976). This trial involved 10 treatments, namely, 5 doses of oral Pirbuterol, 4 doses of Salbutamol and one placebo, applied to 10 patients suffering from bronchial asthma, each being given all 10 treatments over two 5-day weeks. For several other examples of applications of crossover designs in clinical trials and additional references, see Pocock (1983). □

Example 1.1.3. An experiment, reported by Low, Lewis and Prescott (1999) quoting Ballinger, Pickering, Bannister and McLellan (1995), relates to the evaluation of four makes of commode for the use of disabled adults. Suppose the equipments (commodes) are labeled A, B, C and D and a design with four periods (of one week duration each) is to be used. If 8 subjects (disabled adult patients) are available for the experiment, then one possible crossover design is the following.

A	B	C	D	A	B	C	D
B	C	D	A	D	A	B	C
D	A	B	C	B	C	D	A .
C	D	A	B	C	D	A	B

□

Example 1.1.4. An early application of crossover designs in industrial experimentation was given by Williams (1949). We describe an example given by Raghavarao (1990), which relates to a situation in industry where there is a potential for using a crossover design. A manufacturing company with four plants is interested in studying the production rates associated with rotation shifts of different durations. Suppose there are four rotation shifts, namely, weekly (A), biweekly (B), triweekly (C) and monthly (D). Each rotation is to be tried for a period of three months before a change takes place in the rotation. One can use a crossover design of the type shown below for planning such an experiment, where the columns stand for the different plants and the rows represent the periods (first three months, next three months, the three months thereafter, and the last three months):

A	B	C	D
B	C	D	A
D	A	B	C
C	D	A	B

. □

Example 1.1.5. An experiment to assess whether the perceived velocity of a moving point on a computer screen is affected by the presence of vertical or horizontal lines and the spacing between them was described by McNutty (1986). The treatments in this experiment had a factorial structure, there being two factors. The first factor, namely, speed, was at two levels, $S_1 = 2.88$ cm/sec and $S_2 = 6.62$ cm/sec. The other factor was display, having four levels, coded as $D_1, \ldots, D_4$. All possible combinations of the levels of these two factors were taken as treatments. McNutty (1986) had 16 subjects and she used a crossover design given by two replicates of a Williams square of order 8 (for a description of Williams squares, see Section 2.6), and each of the 8 treatment sequences of the square was assigned to two subjects. □

Remark 1.1.1. In the above examples, it is seen that each subject receives distinct treatments over the different periods. However, in some experiments, a subject may also receive the same treatment in more than one period. The following is a crossover design with 3 treatments, 6 periods and 9 subjects. This design allocates each treatment twice to every subject and some of these allocations are done in successive periods. Here the treatments are labeled 1, 2 and 3.

1	1	1	2	2	2	3	3	3
1	2	3	1	2	3	1	2	3
2	2	2	3	3	3	1	1	1
2	3	1	2	3	1	2	3	1
3	3	3	1	1	1	2	2	2
3	1	2	3	1	2	3	1	2

. □

One obvious advantage of using a crossover design is that it requires fewer experimental subjects for the same number of observations. This is an important aspect, especially in situations where the experimental subjects are scarce or expensive. Another advantage of crossover designs is that by a suitable choice of treatment sequences, it is possible to estimate important treatment contrasts within each subject. The precision of such estimates would then be measured by the residual variation between observations on the same subject which is likely to be smaller than the variation between subjects.

The above advantages, however, come at a price that one has to pay while using a crossover design. Compared to a (conventional) design where each experimental subject is exposed to only one treatment, a crossover design (where each subject is exposed to treatments over a number of time periods) might result in a longer duration of the experiment. Therefore, in actual experimentation, designs with a large number of periods may not be particularly attractive to some experimenters. Another major concern is the presence of carryover effects. A distinctive feature of crossover designs is that in any given period, an observation from a subject is affected not only by the *direct* effect of a treatment in the period in which it is applied, but also possibly by the effect of a treatment (or treatments) applied in the preceding period(s) to the same subject. That is, the effect of a treatment might also carry over to one or more of the succeeding periods, giving rise to *residual* or *carryover* effect(s). For instance, in Example 1.1.2, the observation arising out of the site corresponding to the first subject in the second period is influenced by the direct effect of treatment B as also possibly by the residual effect of treatment A. Similarly, the observation arising out of the site corresponding to the third period in the second subject is influenced by the direct effect of treatment A and possibly by the residual effects of the other treatments applied to the same subject in the earlier periods. In particular, an effect carrying over to the immediate next period is referred to as the *first order* carryover effect. Extending this idea of carryover, a treatment applied to a subject in period i has a *second order* carryover effect in period $i + 2$. The generalization to still higher order carryover effects is obvious. Henceforth, we shall use the term carryover effect, instead of residual effect.

The existence of carryover effects in long-term agricultural experiments was first observed by Cochran (1939). Another early example of an experiment indicating the presence of carryover effects was given by Williams (1949). See also John and Quenouille (1977, Chapter 11). The presence of carryover effects often complicates the design and analysis of crossover trials. One way to avoid this difficulty is to insert a *rest* (or *wash out*) period between two successive periods and hope that the carryover effect, if any, would wash out during the rest period. Insertion of rest periods effectively increases the interval between the observed periods and this strategy may help in overcoming the above difficulty if the carryover effect is not expected to persist for a long duration. However, the option of inserting rest periods is not always feasible. One reason is that the insertion of a rest period between each pair of successive periods will certainly increase the total duration of the experiment. Another problem with this option is that there is no guarantee that the rest period is long enough to wash out the carryover effect completely. Furthermore, in the context of clinical trials, insertion of a rest period might invite ethical objections in the sense that a patient cannot be denied a treatment for a long interval just because a rest period has to be inserted. An alternative to the proposal of inserting rest periods is to design the experiment in such a manner that the direct effect differences can be efficiently estimated after adjusting for the presence of possible carryover effects. A large body of the available literature deals with the problem of finding suitable sequences of treatments that allow the estimation of direct effect contrasts in an efficient manner after elimination of carryover effects.

Use of crossover trials, however, is generally limited to situations where the application of a treatment does not significantly alter the subject. For example, in the context of a clinical trial, surgical interventions are unsuitable for crossover trials if the surgery permanently alters the condition. On the other hand, crossover designs in the context of clinical trials are typically useful for experiments on chronic ailments like asthma, migraine, hypertension, epilepsy etc. For some examples of applications in this area, see Pocock (1983).

In spite of some of the above-mentioned problems associated with the use of a crossover design, its advantages make it attractive to an experimenter. As a result, such designs are used in practice quite extensively and the task of their optimal or efficient designing assumes significant practical importance. Though crossover designs have been in use for several decades, the issues relating to the determination of optimal crossover designs have been addressed only in the last 30 years or so. This area of study was pioneered by Hedayat and Afsarinejad (1978) and since then, there has been a continuous flow of work in this area covering optimality aspects under different underlying models and the construction of optimal designs. In this book, we focus on the choice of a "good" or optimal crossover design. While such a choice is dictated by statistical considerations, practical issues also play a role in actual applications. As we shall see presently, the optimality of crossover designs for direct and carryover effects are studied under an assumed model, which at best can be regarded as an approximation to the real relationship between the response and the effects included in the model. Therefore, an optimal design derived under the assumed model might not remain so if the model is mis-specified. In view of this, optimal designs should not always be employed blindly; rather, results on optimal designs in a given context may be used as a step towards selecting good designs and to weed out designs that are indeed poor from a statistical standpoint.

1.2 Notation, Terminology and Models

We begin by listing some notation and terminology in matrix theory. Throughout this book, we deal exclusively with *real* matrices and vectors. All vectors will be written as column vectors, unless specified otherwise, and such vectors will be denoted by boldface numerals or lower case letters. A prime over a matrix or vector will denote its transpose. For a positive integer a, $\mathbf{1}_a$ and I_a will denote the $a \times 1$ vector of all ones and the identity matrix of order a, respectively. For positive integers a and b, $\mathbf{0}_{ab}$ will denote the $a \times b$ null matrix and J_{ab}, the $a \times b$ matrix of all ones, i.e., $J_{ab} = \mathbf{1}_a \mathbf{1}_b'$; J_{aa} will simply be denoted by J_a and similarly, $\mathbf{0}_{a1}$ will be denoted by $\mathbf{0}_a$. Also, for a positive integer a, we will use H_a to denote the matrix $I_a - a^{-1}J_a$, i.e., $H_a = I_a - a^{-1}J_a$. The subscripts will be dropped when there is no confusion regarding the order of the matrices involved.

For a matrix A, A^- will denote an arbitrary generalized inverse (g-inverse) of A, i.e., A^- is a solution of the matrix equation $AXA = A$. Also, for a (real) matrix A, we denote by A^+ the unique Moore-Penrose inverse of A, which satisfies the following conditions:
(i) $AA^+A = A$, (ii)$A^+AA^+ = A^+$, (iii) AA^+ is symmetric and, (iv) A^+A is symmetric.

For a pair of nonnegative definite (n.n.d.) matrices A and B of the same order, we write $A \geq B$ to mean that $A - B$ is n.n.d. and in such a case, we equivalently also write that A majorizes B under the *Loewner ordering* or simply, $A \geq B$ in the Loewner sense.

An $a \times a$ matrix A will be called *completely symmetric* if $A = \alpha I_a + \beta J_a$ for some scalars α and β. The trace of a square matrix A will be denoted by $\mathrm{tr}(A)$.

Consider the partitioned matrix

$$X = \begin{bmatrix} A & B \\ C & D \end{bmatrix},$$

where A and D are non-null square matrices. Also, let $\mathcal{C}(B) \subset \mathcal{C}(A)$, $\mathcal{R}(C) \subset \mathcal{R}(A)$, where $\mathcal{C}(\cdot)$ and $\mathcal{R}(\cdot)$, respectively, denote the column space and row space of a matrix. Then the product CA^-B is invariant with respect to the choice of a g-inverse of A. The (generalized)

Schur complement of A in X is defined to be $D - CA^{-}B$. Similarly, if $\mathcal{C}(C) \subset \mathcal{C}(D)$, $\mathcal{R}(B) \subset \mathcal{R}(D)$, then the (generalized) Schur complement of D in X is $A - BD^{-}C$.

For any matrix A, we write $\text{pr}(A)$ to denote the orthogonal projection matrix onto $\mathcal{C}(A)$, the column space of A. It is well known that $\text{pr}(A)$ is given by $\text{pr}(A) = A(A'A)^{-}A'$. The orthogonal projection matrix onto the space that is orthogonal to $\mathcal{C}(A)$ is denoted by $\text{pr}^{\perp}(A) = I - \text{pr}(A)$, where I is the identity matrix of appropriate order. A basic fact on orthogonal projection matrices will be used later in the book and is stated below as a lemma.

Lemma 1.2.1. *For a partitioned matrix $A = [B \;\; C]$, it can be seen that*

$$\text{pr}(A) = \text{pr}(B) + \text{pr}(\text{pr}^{\perp}(B)C),$$

or equivalently,

$$\text{pr}^{\perp}(A) = \text{pr}^{\perp}(B) - \text{pr}^{\perp}(B)C(C'\text{pr}^{\perp}(B)C)^{-}C'\text{pr}^{\perp}(B).$$

□

Finally, for a pair of matrices E and F, $E \otimes F$ will denote the Kronecker (tensor) product of E and F, i.e., if $E = (e_{ij})$, then $E \otimes F = (e_{ij}F)$. The following are some well known and easily verifiable facts about Kronecker product of matrices:

(a) $(A \otimes B)(C \otimes D) = AC \otimes BD$, provided the products AC and BD are well defined.

(b) $(A_1 + A_2) \otimes B = A_1 \otimes B + A_2 \otimes B$.

(c) $A \otimes (B_1 + B_2) = A \otimes B_1 + A \otimes B_2$.

(d) For scalars a and b, $aA \otimes bB = ab(A \otimes B)$.

(e) $(A \otimes B)' = A' \otimes B'$.

(f) If A, B are invertible matrices, then $(A \otimes B)^{-1} = A^{-1} \otimes B^{-1}$.

Next, we describe some notation and terminology needed for the analysis of a crossover design. Consider a crossover experiment in which $t \geq 2$ treatments are to be compared via n experimental subjects over p time periods. As stated earlier, any allocation of the t treatments to the np experimental sites is called a crossover design. Let $\Omega_{t,n,p}$ be the class of all such crossover designs. For a design $d \in \Omega_{t,n,p}$, $d(i,j)$ denotes the treatment allocated to the jth subject in the ith period according to the design d, $1 \leq i \leq p$, $1 \leq j \leq n$.

We postulate a linear model for the observations generated from a crossover design and then determine optimal designs under a specified model. A variety of models have been considered in the literature. We first consider the following model:

$$\begin{aligned} Y_{ij} = \mu + \alpha_i + \beta_j + \tau_{d(i,j)} + \rho_{d(i-1,j)} + \epsilon_{ij}, \\ 1 \le i \le p, \ \ 1 \le j \le n, \end{aligned} \tag{1.2.1}$$

where Y_{ij} is the observable random variable corresponding to the observation from the jth subject in the ith period, and $\mu, \alpha_i, \beta_j, \tau_s$ and ρ_s are, respectively, a general mean, the ith period effect, the jth subject effect, the direct effect due to treatment s and the first order carryover effect due to treatment s, $1 \le i \le p$, $1 \le j \le n$, $1 \le s \le t$; the ϵ_{ij}'s are the error components, assumed to be uncorrelated random variables with zero means and constant variance σ^2. In the above, we define $\rho_{d(0,j)} = 0$, $1 \le j \le n$. Henceforth, we shall use the same notation Y_{ij} for the observation as well as the random variable corresponding to the observation. This model has traditionally been considered in the literature and so we shall refer to this model as the *traditional* model.

We now make some general comments on the model (1.2.1). To begin with, all the parameters in the model are considered as *fixed*, i.e., non-random. Models with some random effects have been considered in the literature (see, e.g., Mukhopadhyay and Saha (1983), Laska and Meisner (1985), Jones, Kunert and Wynn (1992) and Carriere and Reinsel (1993)) and we elaborate on such models later in this book. It should also be noted that in (1.2.1), only the first order carryover effects have been included. More complex models, incorporating higher order carryover effects have also been considered in the literature and we consider these subsequently in the book. Since $\rho_{d(0,j)} = 0$ for all j, model (1.2.1) has no carryover effects for the observations in the first period. A model of this kind has been referred to in the literature as a *non-circular* model (see, e.g., Hedayat and Afsarinejad (1978) and Cheng and Wu (1980)). An alternative model, called *circular* model, has also been considered by some authors. In a circular model, there are carryover effects for the observations in the first period as well, these arising out of treatments given to the subjects in a pre-period or baseline period. Thus, under a circular model, there are carryover effects in all the p periods and this makes the analysis much simpler compared to that under a non-circular model. Throughout this book, we consider only a non-circular model. For some results on designs under a circular model, see, e.g., Dey and Balachandran (1976), Magda (1980), Kunert (1984a) and Afsarinejad (1989, 1990a). Some authors have also considered simpler models where one or more effects in (1.2.1) are absent. We do not consider such models in this book, except briefly in Chapter 7.

In (1.2.1), a major assumption is that the errors are uncorrelated, having a constant variance. In most studies, the analysis of crossover designs has been carried out under such an assumption of uncorrelated errors. This assumption is however not always reasonable, as observations arising out of the same subject in different periods may be correlated. Issues concerning good crossover designs when the observations coming from the same subject are possibly correlated have been examined in the literature and we shall return to this aspect in Chapters 4, 6 and 7, where relevant references may be found.

Again, in (1.2.1), it has been assumed that the carryover effect of a treatment is the same irrespective of which treatment follows it in the next period on the same subject. However, in some experimental situations, in a given period, the carryover effect may depend on the treatment that contributes the direct effect. For example, if A is followed by A in two consecutive periods, then the carryover effect in the second period is different from that when A is followed by B. If a treatment is followed by itself, the carryover effect in the next period is called a self carryover effect while carryover effects contributed by other treatments are termed as mixed carryover effects. Models that account for such carryover effects have been studied recently and we deal with such models in Chapter 5.

Another modification of the traditional model (1.2.1) is one in which the carryover effect of a treatment is assumed to be proportional to its direct effect as it seems reasonable to assume that a treatment with large direct effect should generally have a large carryover effect. Optimal designs under such a model have been considered in Chapter 5.

Yet another modification of (1.2.1) is one which incorporates interactions between the treatments contributing the direct and carryover effects in two successive periods on the same subject together with their individual direct and carryover effects. Such a non-additive model and related optimal designs are reviewed in Chapter 6.

Several construction methods described in this book require a basic knowledge of finite groups and finite fields. One may refer, e.g., to Jacobson (1964) for a description of these aspects. Reference is also made to Dean and Voss (1999) for an excellent exposition of the general area of experimental designs.

1.3 Information Matrices

In order to present the various optimality results, we need to introduce some further notation which will be used throughout the book. Recall that $\Omega_{t,n,p}$ is the class of all crossover designs involving t treatments applied to n subjects over p periods. For $d \in \Omega_{t,n,p}$ and $1 \leq s, s' \leq t$, $1 \leq i \leq p$, $1 \leq j \leq n$, let

n_{dsj} = number of times treatment s is allocated to subject j,
$\bar{n}_{dsj}$ = number of times treatment s is allocated to subject j in the first $p-1$ periods,
m_{dsi} = number of times treatment s is allocated to period i,
$z_{dss'}$ = number of times treatment s is immediately preceded by treatment s' on the same subject,
r_{ds} = number of times treatment s appears in the design,
$\bar{r}_{ds}$ = number of times treatment s appears in the first $p-1$ periods.

Again, for $d \in \Omega_{t,n,p}$, using the terms defined above, let

$$\begin{aligned}
\boldsymbol{r}_d &= (r_{d1}, \ldots, r_{dt})', & \bar{\boldsymbol{r}}_d &= (\bar{r}_{d1}, \ldots, \bar{r}_{dt})', \\
R_d &= \mathrm{diag}(r_{d1}, \ldots, r_{dt}), & \bar{R}_d &= \mathrm{diag}(\bar{r}_{d1}, \ldots, \bar{r}_{dt}), \\
N_d &= (n_{dsj})_{\substack{1\leq s\leq t;\\ 1\leq j\leq n}}, & \bar{N}_d &= (\bar{n}_{dsj})_{\substack{1\leq s\leq t;\\ 1\leq j\leq n}}, \\
M_d &= (m_{dsi})_{\substack{1\leq s\leq t;\\ 1\leq i\leq p}}, & \bar{M}_d &= \left[\boldsymbol{0}_t \;\; (m_{dsi})_{\substack{1\leq s\leq t;\\ 1\leq i\leq p-1}}\right] \\
Z_d &= (z_{dss'})_{1\leq s,s'\leq t}.
\end{aligned} \tag{1.3.1}$$

Thus $\boldsymbol{r}_d$ (respectively, $\bar{\boldsymbol{r}}_d$) is the replication vector for direct (respectively, carryover) effects; N_d (respectively, $\bar{N}_d$) is the $t \times n$ direct (respectively, carryover) effect versus subject incidence matrix; M_d (respectively, $\bar{M}_d$) is the $t \times p$ direct (respectively, carryover) effect versus period incidence matrix and Z_d is the $t \times t$ direct effect versus carryover effect incidence matrix. It is easy to verify that

$$\begin{aligned}
\boldsymbol{r}_d'\boldsymbol{1}_t &= np, \; \bar{\boldsymbol{r}}_d'\boldsymbol{1}_t = n(p-1), \\
N_d\boldsymbol{1}_n &= \boldsymbol{r}_d, \; \boldsymbol{1}_t'N_d = p\boldsymbol{1}_n', \\
\bar{N}_d\boldsymbol{1}_n &= \bar{\boldsymbol{r}}_d, \; \boldsymbol{1}_t'\bar{N}_d = (p-1)\boldsymbol{1}_n', \\
M_d\boldsymbol{1}_p &= \boldsymbol{r}_d, \; \boldsymbol{1}_t'M_d = n\boldsymbol{1}_p', \\
\bar{M}_d\boldsymbol{1}_p &= \bar{\boldsymbol{r}}_d, \; \boldsymbol{1}_t'\bar{M}_d = (0, n\boldsymbol{1}_{p-1}') \\
\boldsymbol{1}_t'Z_d &= \bar{\boldsymbol{r}}_d'.
\end{aligned} \tag{1.3.2}$$

It is useful to write model (1.2.1) in matrix form and we first develop suitable notation for doing this. For any design $d \in \Omega_{t,n,p}$, let

$$\boldsymbol{Y}_d = (Y_{11}, \ldots, Y_{p1}, Y_{12}, \ldots, Y_{p2}, \ldots, Y_{1n}, \ldots, Y_{pn})'$$

be the $np \times 1$ vector of observations arising out of d with Y_{ij} as in (1.2.1). This simply means that the first p entries in $\boldsymbol{Y}_d$ correspond to the p observations on subject 1, the next p to the observations on subject 2, ..., the last p pertaining to the observations on subject n. Also, let $\boldsymbol{\alpha} = (\alpha_1, \ldots, \alpha_p)'$, $\boldsymbol{\beta} = (\beta_1, \ldots, \beta_n)'$, $\boldsymbol{\tau} = (\tau_1, \ldots, \tau_t)'$, $\boldsymbol{\rho} = (\rho_1, \ldots, \rho_t)'$ and $\boldsymbol{\epsilon} = (\epsilon_{11}, \ldots, \epsilon_{pn})'$ be the $p \times 1$ vector of period effects, the $n \times 1$ vector of subject effects, the $t \times 1$ vector of direct effects, the $t \times 1$ vector of carryover effects and the $np \times 1$ vector of error terms, respectively, where $\alpha_i, \beta_j, \tau_s, \rho_s$ and ϵ_{ij} are as in (1.2.1). Also, let $\boldsymbol{\theta} = (\mu, \boldsymbol{\alpha}', \boldsymbol{\beta}', \boldsymbol{\tau}', \boldsymbol{\rho}')'$ with μ as in (1.2.1).

For a design $d \in \Omega_{t,n,p}$, let T_{dj} be a $p \times t$ matrix with its (i, s)th entry equal to 1 if subject j receives the direct effect of treatment s in the ith period, and zero otherwise. Similarly, let F_{dj} be a $p \times t$ matrix with its (i, s)th entry equal to 1 if subject j has the carryover effect of treatment s in the ith period, and zero otherwise. Thus, the first row of F_{dj} is zero and for $2 \le i \le p, 1 \le j \le n$, the ith row of F_{dj} is the $(i-1)$th row of T_{dj}, i.e., $F_{dj} = \begin{pmatrix} \mathbf{0}'_{p-1} & 0 \\ I_{p-1} & \mathbf{0}_{p-1} \end{pmatrix} T_{dj}$, $1 \le j \le n$. Define

$$T_d = (T'_{d1}, \ldots, T'_{dn})' \text{ and } F_d = (F'_{d1}, \ldots, F'_{dn})'.$$

Let $\mathbb{E}(\cdot)$ and $\mathbb{D}(\cdot)$ respectively, denote the expectation and dispersion operators.

With the above notation, model (1.2.1) can equivalently be written in matrix form as

$$\begin{aligned} \boldsymbol{Y}_d &= X_d\boldsymbol{\theta} + \boldsymbol{\epsilon}, \\ \mathbb{E}(\boldsymbol{\epsilon}) &= \mathbf{0}, \quad \mathbb{D}(\boldsymbol{\epsilon}) = \sigma^2 I_{np}, \end{aligned} \tag{1.3.3}$$

where the design matrix X_d may be written in the following partitioned form

$$X_d = [\mathbf{1}_{np} \quad P \quad U \quad T_d \quad F_d], \tag{1.3.4}$$

with P, U, T_d and F_d as the parts of the design matrix corresponding to the period, subject, direct and carryover effects, respectively, under the design d. The orders of P, U, T_d and F_d are, respectively, $np \times p$, $np \times n$, $np \times t$ and $np \times t$. Furthermore, with the ordering of the observations as in $\boldsymbol{Y}_d$, it is clear that

$$P = \mathbf{1}_n \otimes I_p \text{ and } U = I_n \otimes \mathbf{1}_p.$$

It can readily be seen that

$$\begin{array}{llll} T_d\mathbf{1}_t &= \mathbf{1}_{np}, & \mathbf{1}'_{np}T_d = & \boldsymbol{r}'_d, \\ F_d\mathbf{1}_t &= \mathbf{1}_n \otimes \left(0, \mathbf{1}'_{p-1}\right)', & \mathbf{1}'_{np}F_d = & \bar{\boldsymbol{r}}'_d, \\ P\mathbf{1}_p &= \mathbf{1}_{np}, & \mathbf{1}'_{np}P = & n\mathbf{1}'_p, \\ U\mathbf{1}_n &= \mathbf{1}_{np}, & \mathbf{1}'_{np}U = & p\mathbf{1}'_n, \\ P'P &= nI_p, & U'U = & pI_n, \\ PP' &= J_n \otimes I_p, & UU' = & I_n \otimes J_p, \\ P'U &= J_{pn}, & P'T_d = & M'_d \\ P'F_d &= \bar{M}'_d, & F'_dF_d = & \bar{R}_d, \\ U'T_d &= N'_d, & U'F_d = & \bar{N}'_d, \\ T'_dT_d &= R_d, & T'_dF_d = & Z_d. \end{array} \tag{1.3.5}$$

Now, in order to obtain the information matrices for direct and carryover effects under model (1.3.3), one needs to start with the matrix X'_dX_d where X_d is as in (1.3.4). Remembering that $\boldsymbol{\alpha}$ and $\boldsymbol{\beta}$ are the nuisance parameters, we may further partition X_d as $X_d = [\mathbf{1}_{np} \quad X_1 \quad X_2]$, with

$$X_1 = [P \;\; U], \quad X_2 = [T_d \;\; F_d]. \tag{1.3.6}$$

Then

$$X'_dX_d = \begin{bmatrix} \mathbf{1}'_{np}\mathbf{1}_{np} & \mathbf{1}'_{np}X_1 & \mathbf{1}'_{np}X_2 \\ X'_1\mathbf{1}_{np} & X'_1X_1 & X'_1X_2 \\ X'_2\mathbf{1}_{np} & X'_2X_1 & X'_2X_2 \end{bmatrix} = \begin{bmatrix} np & n\mathbf{1}'_p & p\mathbf{1}'_n & \boldsymbol{r}'_d & \bar{\boldsymbol{r}}'_d \\ n\mathbf{1}_p & nI_p & J_{pn} & M'_d & \bar{M}'_d \\ p\mathbf{1}_n & J_{np} & pI_n & N'_d & \bar{N}'_d \\ \boldsymbol{r}_d & M_d & N_d & R_d & Z_d \\ \bar{\boldsymbol{r}}_d & \bar{M}_d & \bar{N}_d & Z'_d & \bar{R}_d \end{bmatrix}, \tag{1.3.7}$$

on simplification, using (1.3.5) and (1.3.6).

For a design $d \in \Omega_{t,n,p}$, let C_d and $\bar{C}_d$, respectively, denote the information matrix (the so-called C-matrix) of the direct and carryover effects, i.e., C_d (respectively, $\bar{C}_d$) is the coefficient matrix in the reduced normal equations for $\boldsymbol{\tau}$ (respectively, $\boldsymbol{\rho}$). Both C_d and $\bar{C}_d$ are $t \times t$ symmetric matrices. A crossover design is said to be *connected* for direct (respectively, carryover) effects if all contrasts among the direct (respectively, carryover) effects are estimable. Clearly, a necessary and sufficient condition for a crossover design to be connected for direct (respectively, carryover) effects is that $\text{Rank}(C_d) = t - 1$ (respectively, $\text{Rank}(\bar{C}_d) = t - 1$).

From (1.3.5), $P\mathbf{1}_p = U\mathbf{1}_n = \mathbf{1}_{np}$, i.e., $\mathbf{1}_{np}$ belongs to the column space of X_1. Hence it follows from (1.3.7) that under model (1.3.3) the information matrix for estimating $\boldsymbol{\tau}$ and $\boldsymbol{\rho}$ jointly is of the form

$$C_d(\boldsymbol{\tau},\boldsymbol{\rho}) = X_2'X_2 - X_2'X_1(X_1'X_1)^- X_1'X_2.$$

Noting that a choice of a g-inverse of $X_1'X_1$ is given by

$$(X_1'X_1)^- = \begin{pmatrix} nI_p & J_{pn} \\ J_{np} & pI_n \end{pmatrix}^- = \begin{pmatrix} n^{-1}I_p & \mathbf{0}_{pn} \\ -(np)^{-1}J_{np} & p^{-1}I_n \end{pmatrix},$$

$C_d(\boldsymbol{\tau},\boldsymbol{\rho})$ above may be simplified to

$$C_d(\boldsymbol{\tau},\boldsymbol{\rho}) = X_2'AX_2 = \begin{bmatrix} T_d'AT_d & T_d'AF_d \\ F_d'AT_d & F_d'AF_d \end{bmatrix}, \tag{1.3.8}$$

where $A = (I_n - n^{-1}J_n) \otimes (I_p - p^{-1}J_p) = H_n \otimes H_p$, writing as before, $H_a = I_a - a^{-1}J_a$ for a positive integer a. Consequently, using (1.3.6) and (1.3.7), we may write

$$C_d(\boldsymbol{\tau},\boldsymbol{\rho}) = \begin{bmatrix} C_{d11} \; C_{d12} \\ C_{d21} \; C_{d22} \end{bmatrix}, \tag{1.3.9}$$

where

$$\begin{aligned} C_{d11} &= R_d - n^{-1}M_dM_d' - p^{-1}N_dN_d' + (np)^{-1}\boldsymbol{r}_d\boldsymbol{r}_d', \\ C_{d12} &= Z_d - n^{-1}M_d\bar{M}_d' - p^{-1}N_d\bar{N}_d' + (np)^{-1}\boldsymbol{r}_d\bar{\boldsymbol{r}}_d' = C_{d21}', \\ C_{d22} &= \bar{R}_d - n^{-1}\bar{M}_d\bar{M}_d' - p^{-1}\bar{N}_d\bar{N}_d' + (np)^{-1}\bar{\boldsymbol{r}}_d\bar{\boldsymbol{r}}_d'. \end{aligned} \tag{1.3.10}$$

Hence, from (1.3.9), under model (1.3.3), C_d and $\bar{C}_d$, the information matrices for direct effects and carryover effects are, respectively, given by the Schur complements of C_{d22} and C_{d11} in $C_d(\boldsymbol{\tau},\boldsymbol{\rho})$. Thus we have

$$\begin{aligned} C_d &= C_{d11} - C_{d12}C_{d22}^-C_{d21} \\ \bar{C}_d &= C_{d22} - C_{d21}C_{d11}^-C_{d12}. \end{aligned} \tag{1.3.11}$$

It can be shown that C_d and $\bar{C}_d$ as in (1.3.11) are invariant with respect to the choice of g-inverses involved. Equivalently, the matrices $C_{duv}, u, v = 1, 2$, in (1.3.10) can be written using projection operators as

$$\begin{aligned} C_{d11} &= T_d'\mathrm{pr}^{\perp}([P \;\; U])\,T_d, \\ C_{d12} &= T_d'\mathrm{pr}^{\perp}([P \;\; U])\,F_d = C_{d21}', \\ C_{d22} &= F_d'\mathrm{pr}^{\perp}([P \;\; U])\,F_d. \end{aligned} \tag{1.3.12}$$

Similarly, using projection operators we may directly write down expressions for C_d and $\bar{C}_d$ from model (1.3.3) with X_d as in (1.3.4). By Lemma 1.2.1, these expressions are clearly equivalent to those in (1.3.11) and are given by

$$\begin{aligned} C_d &= T_d'\mathrm{pr}^{\perp}([\mathbf{1}_{np} \;\; P \;\; U \;\; F_d])T_d \\ &= T_d'\mathrm{pr}^{\perp}([P \; U \; F_d])T_d \\ &= T_d'\mathrm{pr}^{\perp}([P \;\; U])T_d - T_d'\mathrm{pr}(\mathrm{pr}^{\perp}([P \; U])F_d)T_d, \\ \bar{C}_d &= F_d'\mathrm{pr}^{\perp}([\mathbf{1}_{np} \;\; P \;\; U \;\; T_d])F_d \\ &= F_d'\mathrm{pr}^{\perp}([P \; U \; T_d])F_d \\ &= F_d'\mathrm{pr}^{\perp}([P \;\; U])F_d - F_d'\mathrm{pr}(\mathrm{pr}^{\perp}([P \;\; U])T_d)F_d. \end{aligned} \quad (1.3.13)$$

As noted previously, $\mathbf{1}_{np}$ belongs to the column space of $[P \; U]$ and hence to the columns spaces of $[P \; U \; F_d]$ as well as $[P \; U \; T_d]$. These facts have been used in obtaining (1.3.12) and (1.3.13). Furthermore, from (1.3.5), both $T_d\mathbf{1}_t = \mathbf{1}_n \otimes \mathbf{1}_p$ and $F_d\mathbf{1}_t = \mathbf{1}_n \otimes (0, \; \mathbf{1}_{p-1}')'$ belong to the column space of $P = \mathbf{1}_n \otimes I_p$. Hence by (1.3.13), $C_d\mathbf{1}_t = \bar{C}_d\mathbf{1}_t = \mathbf{0}$. Since both C_d and $\bar{C}_d$ are symmetric, it is clear that both the matrices have zero row and column sums for every $d \in \Omega_{t,n,p}$.

In the above development we have assumed that all the errors are uncorrelated. This assumption may become untenable if we believe that the observations arising from the same subject are correlated while those from different subjects remain uncorrelated. For such situations we may assume that the errors $\boldsymbol{\epsilon}_j = (\epsilon_{1j}, \ldots, \epsilon_{pj})'$ are independent while $D(\boldsymbol{\epsilon}_j) = V$ for all j, $1 \le j \le n$, where V is a $p \times p$ positive definite matrix. This assumption modifies model (1.3.3) to

$$\boldsymbol{Y}_d = X_d\boldsymbol{\theta} + \boldsymbol{\epsilon}, \;\; \mathbb{E}(\boldsymbol{\epsilon}) = \mathbf{0}, \;\; \mathbb{D}(\boldsymbol{\epsilon}) = \sigma^2\Sigma, \quad (1.3.14)$$

where

$$\Sigma = I_n \otimes V$$

and other terms are as in (1.3.3). When $V = I_p$, model (1.3.14) reduces to model (1.3.3).

Under model (1.3.14), we consider the weighted least squares estimates and so, instead of $X_d'X_d$, we need to consider the matrix $X_d'\Sigma^{-1}X_d$, where

$$\Sigma^{-1} = I_n \otimes V^{-1}.$$

Consequently, as before, the information matrix for estimating $\boldsymbol{\tau}$ and $\boldsymbol{\rho}$ jointly under model (1.3.14) is of the form

$$C_d(\boldsymbol{\tau}, \boldsymbol{\rho}) = X_2'\Sigma^{-1}X_2 - X_2'\Sigma^{-1}X_1(X_1'\Sigma^{-1}X_1)^{-}X_1'\Sigma^{-1}X_2, \quad (1.3.15)$$

where X_1, X_2 are as in (1.3.6). On simplification using the forms of P and U, we have

$$X_1'\Sigma^{-1}X_1 = \begin{bmatrix} nV^{-1} & \mathbf{1}_n' \otimes V^{-1}\mathbf{1}_p \\ \mathbf{1}_n \otimes (\mathbf{1}_p'V^{-1}) & (\mathbf{1}_p'V^{-1}\mathbf{1}_p)I_n \end{bmatrix}.$$

Writing $\delta = (\mathbf{1}_p'V^{-1}\mathbf{1}_p)^{-1}$, we note that

$$(X_1'\Sigma^{-1}X_1)\begin{bmatrix} n^{-1}(\mathbf{1}_n' \otimes I_p) \\ \delta H_n \otimes (\mathbf{1}_p'V^{-1}) \end{bmatrix} = \begin{bmatrix} (\mathbf{1}_n' \otimes V^{-1}) \\ I_n \otimes (\mathbf{1}_p'V^{-1}) \end{bmatrix} = X_1'\Sigma^{-1}$$

$$\Rightarrow (X_1'\Sigma^{-1}X_1)^- X_1'\Sigma^{-1} = \begin{bmatrix} n^{-1}(\mathbf{1}_n' \otimes I_p) \\ \delta H_n \otimes (\mathbf{1}_p'V^{-1}) \end{bmatrix}.$$

Using this in (1.3.15), upon simplification, we get that under model (1.3.14),

$$C_d(\boldsymbol{\tau}, \boldsymbol{\rho}) = X_2'A^*X_2 = \begin{bmatrix} C_{d11} & C_{d12} \\ C_{d21} & C_{d22} \end{bmatrix} = \begin{bmatrix} T_d'A^*T_d & T_d'A^*F_d \\ F_d'A^*T_d & F_d'A^*F_d \end{bmatrix}, \quad (1.3.16)$$

where $A^* = H_n \otimes \left(V^{-1} - \delta V^{-1}J_pV^{-1}\right) = H_n \otimes V^*$, writing

$$V^* = V^{-1} - \delta V^{-1}J_pV^{-1}. \quad (1.3.17)$$

Now the information matrices C_d and $\bar{C}_d$ under model (1.3.14) may be obtained from (1.3.16) as in (1.3.11). Alternatively, as in (1.3.13), we may directly write equivalent expressions for C_d and $\bar{C}_d$ under model (1.3.14) starting from $X_d'\Sigma^{-1}X_d$, with X_d as in (1.3.4), e.g.,

$$C_d = T_d'(I_n \otimes V^{-\frac{1}{2}})\mathrm{pr}^{\perp}\{(I_n \otimes V^{-\frac{1}{2}})[P\ U\ F_d]\}(I_n \otimes V^{-\frac{1}{2}})T_d, \quad (1.3.18)$$

where $V^{-\frac{1}{2}}$ is a $p \times p$ positive definite matrix such that

$$V^{-\frac{1}{2}}V^{-\frac{1}{2}} = V^{-1}.$$

Arguments similar to those used in the context of uncorrelated errors show that C_d as in (1.3.18) has row and column sums zero and can be written as

$$C_d = H_t(T_d'(I_n \otimes V^{-\frac{1}{2}})\mathrm{pr}^{\perp}\{(I_n \otimes V^{-\frac{1}{2}})[P\ U\ F_dH_t]\}(I_n \otimes V^{-\frac{1}{2}})T_d)H_t.$$

It is interesting to see that when $p = 2$, designs which are optimal under model (1.3.3) remain so even under model (1.3.14). This follows from the following lemma.

Lemma 1.3.1. *When $p = 2$, the two forms of the matrix $C_d(\boldsymbol{\tau}, \boldsymbol{\rho})$ as given by (1.3.8) and (1.3.16) are proportional to each other.*

Proof. Let $\boldsymbol{x}$ be a $p \times 1$ non-null vector. Then with V^* as in (1.3.17),

$$V^*\boldsymbol{x} = V^{-1}\boldsymbol{x} - \delta V^{-1}J_pV^{-1}\boldsymbol{x} = V^{-1}\boldsymbol{x} - \left(\delta \mathbf{1}_p'V^{-1}\boldsymbol{x}\right)V^{-1}\mathbf{1}_p.$$

It follows that

$$\begin{aligned} V^*\boldsymbol{x} = &\ \mathbf{0}, \text{ if and only if} \\ \boldsymbol{x} &= \delta\mathbf{1}_p'V^{-1}\boldsymbol{x}\mathbf{1}_p = \alpha\mathbf{1}_p, \end{aligned} \tag{1.3.19}$$

where α is a constant. For $p = 2$, this implies that the columns of V^* are spanned by $(1, \ -1)'$, and so for $p = 2$, $V^* = h\begin{bmatrix} 1 & -1 \\ -1 & 1 \end{bmatrix} = 2h(I_2 - \frac{1}{2}J_2) = 2hH_2$, where $h > 0$ is some constant involving the elements of V.

Hence from (1.3.16), under model (1.3.14), for $p = 2$,

$$C_d(\boldsymbol{\tau}, \boldsymbol{\rho}) = 2h\left[X_2'\left(H_n \otimes H_2\right)X_2\right].$$

The first factor on the right above, i.e., the constant $2h$, absorbs the correlation while the second factor is a matrix which does not involve the correlation and is identical to the expression obtained from (1.3.8) by putting $p = 2$. Hence the lemma. □

From Lemma 1.3.1, it is clear that for $p = 2$, if the two observations within a subject are correlated then the information matrix C_d is proportional to the information matrix which would have been obtained had the observations been all uncorrelated, and the proportionality constant is a function of the correlation term. Hence if $p = 2$ and one wants to find an optimal design for $\boldsymbol{\tau}$ under model (1.3.14), then it suffices to do so under model (1.3.3). Thus, for the problem of finding an optimal design for $\boldsymbol{\tau}$ when $p = 2$, both these models are equivalent.

In so far as the direct effects are concerned, the above assertion also follows by noting that with $p = 2$, the best linear unbiased estimators of all direct effect contrasts must necessarily be based on only the difference of the two observations within each subject. This is intuitively clear since consideration of such differences is the only way for eliminating the fixed effects due to subjects.

The information matrices for more complex models than (1.3.3) or (1.3.14) may be derived on similar lines, using heavier notation. Such information matrices will be derived in subsequent chapters, where such models are considered.

1.4 Optimality Criteria and Tools

Recall that a crossover design is an allocation of t treatments to the np experimental sites, where n is the number of subjects, each one of which is studied for p periods. Clearly, the total number of possible treatment sequences of length p is t^p. Suppose $\mathcal{S}$ is the collection of these t^p treatment sequences. One might then look upon a crossover design involving n subjects as one which assigns $n_{\boldsymbol{s}}$ subjects to the treatment sequence $\boldsymbol{s} \in \mathcal{S}$, where $n_{\boldsymbol{s}} \geq 0$ and $\sum_{\boldsymbol{s} \in \mathcal{S}} n_{\boldsymbol{s}} = n$.

For given values of the design parameters, t, n and p, typically there will be many choices for the design and these alternatives form a class of competing designs. To discriminate among different designs belonging to a competing class, under a suitable model postulated for the observations generated by them, one needs to compare the designs using some well-defined criterion which depends on the objective of the study. For instance, if comparison of the direct effects of the treatments is of primary interest, then one needs to find a design which is the best, in the competing class and under the chosen criterion, with regard to the estimation of direct effect contrasts. If such a design can be identified, then it is called *optimal* for direct effects in the competing class, with reference to the model and the given criterion. Similarly, optimal designs for carryover effects can be considered.

A systematic study of various optimality criteria (including the popular A-, D- and E-criteria) and related optimal designs in a very broad context was initiated by Kiefer (1958). Another useful optimality criterion, called the MV-optimality, which we use in Chapter 7, was originally introduced by Takeuchi (1961) and the term "MV-optimality" is due to Jacroux (1983). We refer the reader to Shah and Sinha (1989) and Pukelsheim (1993) for excellent accounts of various optimality criteria and optimal designs. Pukelsheim (1993) dealt mostly with what is known as *approximate theory*, where a design is visualized as a probability measure over the experimental region, the probabilities representing the proportion of observations at different sites. The general problem then is to determine these proportions in an "optimal" manner. The *exact theory* on the other hand is concerned with the problem of finding an optimal design for a given finite number of observations. For a more recent and authoritative review of optimal designs (including crossover designs) based on exact theory, we refer to Cheng (1996).

The notion of *universal optimality* helps in unifying various optimality criteria. In particular, it is well known that a universally optimal design is also optimal according to the A-, D-, E- and MV-criteria. In this book, we shall primarily be concerned with the universal optimality criteria and so, brief details on this seem to be in order. Let $\mathcal{A}$ denote the class of symmetric, nonnegative definite (n.n.d.) matrices of order t with zero row sums. Consider the class Φ of real-valued functions $\phi(\cdot)$ defined on $\mathcal{A}$, such that

(a) $\phi(\cdot)$ is convex; that is, for every $A_1, A_2 \in \mathcal{A}$ and real a $(0 \leq a \leq 1)$,

$$a\phi(A_1) + (1-a)\phi(A_2) \geq \phi(aA_1 + (1-a)A_2);$$

(b) $\phi(bA)$ is non-increasing in the scalar $b \geq 0$ for all $A \in \mathcal{A}$;
(c) $\phi(\cdot)$ is permutation invariant, i.e., $\phi(EAE') = \phi(A)$ for every permutation matrix E of order t and for each $A \in \mathcal{A}$.

Let $\mathcal{D}$ be a class of competing designs in a given context and let A_d denote the information matrix for a set of relevant parametric functions (for instance, these may concern the direct or carryover effects in our setup) under a design $d \in \mathcal{D}$ and a given model. Then, under this model, a design $d^* \in \mathcal{D}$ is said to be universally optimal over $\mathcal{D}$ for the estimation of this set of parametric functions if $\phi(A_{d^*}) = \min_{d \in \mathcal{D}} \phi(A_d)$ for every $\phi(\cdot) \in \Phi$. Incidentally, as will be noted later in Section 2.2, in our context $A_d \in \mathcal{A}$ and so $\phi(A_d)$ is well defined for every $\phi(\cdot) \in \Phi$ and every competing design d.

A simple sufficient condition for a design to be universally optimal was provided by Kiefer (1975). Since we shall often be referring to this sufficient condition, this result is stated below in the form of a theorem.

Theorem 1.4.1. *Suppose for a design $d^* \in \mathcal{D}$,*
(i) A_{d^*} is completely symmetric, and,
(ii) $tr(A_{d^*}) \geq \mathrm{tr}(A_d)$ for all $d \in \mathcal{D}$.
Then, under the assumed model and for the set of parametric functions under study, d^* is universally optimal over $\mathcal{D}$. □

As noted by Kiefer (1975), $-\mathrm{tr}(A)$ is a functional satisfying the conditions (a)–(c) above. Hence, a *necessary* condition for a design to be universally optimal is that the corresponding information matrix A has maximum trace among all competing information matrices.

An extension of universal optimality criterion of Kiefer (1975) was considered by Shah and Sinha (2002). Let A_d denote the information matrix for the relevant parametric functions under a suitable model using the design d. Let g be a permutation of $\{1, 2, \ldots, t\}$, that is $g \in S_t$, the symmetric group of permutations on $\{1, 2, \ldots, t\}$. As per Shah and Sinha (2002), a design d^* with information matrix C_{d^*} is said to be universally optimal in an appropriate class of competing designs if it minimizes every real-valued optimality functional $\phi(\cdot)$ defined over the set of n.n.d. matrices, that satisfies the following conditions:

(i) $\phi(A_{d_g}) = \phi(A_d)$ for every $g \in S_t$, where d_g is the design obtained by permuting the treatment labels according to g;
(ii) $A_d \geq A_f \Rightarrow \phi(A_d) \leq \phi(A_f)$, where d and f are any two designs in the competing class;
(iii) $\phi(\sum w_g A_{d_g}) \leq \phi(A_d)$, where $\{w_g\}$ are nonnegative rationals satisfying $\sum_g w_g = 1$. Here g runs over all the $t!$ permutations in S_t.

Note that every convex functional satisfies (iii). This formulation of universal optimality is an extension of the original formulation of Kiefer (1975) in the sense that the condition of convexity in the original formulation is replaced by a slightly weaker condition (iii) above.

A sufficient condition for d_0 to be universally optimal (as per the extended definition) is that

$$\sum w_g A_{d_g} \leq A_{d_0} \quad \text{for every } d. \tag{1.4.1}$$

In (1.4.1), the $\{w_g\}$ can be any specific set of weights (which may depend on d). While using (1.4.1) subsequently in this book, we shall take $w_g = 1/t!$ for all $g \in S_t$.

We now discuss a technique often used for obtaining universally optimal designs in a class $\mathcal{D}$ under a given model. This involves starting with a universally optimal design under a simpler model and then checking a suitable orthogonality condition which, when satisfied, leads to equality of the information matrices under the given model and the simpler model. Consequently, if the orthogonality condition holds, then a universally optimal design under a simpler model remains universally optimal under the given model. This technique was first used by Magda (1980) and then elegantly and extensively studied by Kunert (1983). This approach has been used widely in the search of optimal crossover designs and several theorems in the subsequent chapters make use of this technique. A few details of this technique are summarized below.

Consider the model

$$\boldsymbol{Y}_d = X_{1d}\boldsymbol{\nu}_1 + X_{2d}\boldsymbol{\nu}_2 + \boldsymbol{\epsilon}, \tag{1.4.2}$$

where $\boldsymbol{Y}_d$ is the vector of observations under the design d, $\boldsymbol{\nu}_1$ and $\boldsymbol{\nu}_2$ are two vectors of parameters, X_{1d} and X_{2d} are the corresponding design matrices and $\boldsymbol{\epsilon}$ is the vector of error terms. Suppose $\boldsymbol{\nu}_2 = (\boldsymbol{\nu}_{21}' \; \boldsymbol{\nu}_{22}')'$ and X_{2d} is conformally partitioned as $X_{2d} = [X_{21d} \;\; X_{22d}]$. Then consider the model

$$\boldsymbol{Y}_d = X_{1d}\boldsymbol{\nu}_1 + X_{21d}\boldsymbol{\nu}_{21} + \boldsymbol{\epsilon}. \tag{1.4.3}$$

Following Kunert (1983), we will call model (1.4.2) "finer" than model (1.4.3), while (1.4.3) will be said to be "simpler" than (1.4.2). The information matrices for the estimation of $\boldsymbol{\nu}_1$ under (1.4.2) and (1.4.3) are, respectively, given by

$$A_d = X'_{1d}\text{pr}^{\perp}(X_{2d})X_{1d}, \text{ and } B_d = X'_{1d}\text{pr}^{\perp}(X_{21d})X_{1d}.$$

From Lemma 1.2.1,

$$\text{pr}(X_{2d}) = \text{pr}(X_{21d}) + \text{pr}(\text{pr}^{\perp}(X_{21d})X_{22d}), \tag{1.4.4}$$

and hence

$$\begin{aligned} \text{pr}^{\perp}(X_{2d}) &= \text{pr}^{\perp}(X_{21d}) - \text{pr}\{\text{pr}^{\perp}(X_{21d})X_{22d}\} \\ &\leq \text{pr}^{\perp}(X_{21d}). \end{aligned} \tag{1.4.5}$$

Therefore

$$A_d \leq B_d, \quad \text{for all } d \in \mathcal{D}, \tag{1.4.6}$$

with equality if and only if the following *orthogonality* condition holds:

$$X'_{1d}\text{pr}^{\perp}\left(X_{21d}\right)X_{22d} = \mathbf{0}. \tag{1.4.7}$$

In view of the above, we get the next lemma which is useful in finding optimal designs.

Lemma 1.4.1. *Let $d^* \in \Omega_{t,n,p}$ be such that (i) B_{d^*} is completely symmetric, (ii) $\text{tr}(B_{d^*}) \geq \text{tr}(B_d)\ \forall\ d \in \Omega_{t,n,p}$ and (iii)$X'_{1d^*}\text{pr}^{\perp}(X_{21d^*})X_{22d^*} = \mathbf{0}$. Then d^* is universally optimal for the estimation of $\boldsymbol{\nu}_1$ under (1.4.2).* □

Kushner (1997b) used an argument of Pukelsheim (2003, p. 75) to obtain an upper bound for the trace of the information matrix C_d. This result was further generalized by Kunert and Martin (2000a) in the context of interference models. These results have been used by several authors to identify universally optimal designs under various situations and we shall refer to these in later chapters. Here, we state the result of Kushner (1997b).

Lemma 1.4.2. *For $d \in \Omega_{t,n,p}$, writing $C_d(\boldsymbol{\tau}, \boldsymbol{\rho}) = \begin{bmatrix} C_{d11} & C_{d12} \\ C_{d21} & C_{d22} \end{bmatrix}$ and with $C_d = C_{d11} - C_{d12}C_{d22}^{-}C_{d21}$, an upper bound for $\text{tr}(C_d)$ is given by*

$$\text{tr}(C_d) \leq \text{tr}(C_{d11}) - (\text{tr}(C_{d12}))^2/\text{tr}(C_{d22}).$$

□

1.5 Outline of the Book

Work on the optimality aspects of crossover designs was initiated by Hedayat and Afsarinejad (1978). Since then, research in this area has seen extensive growth and many important results have been proved in recent years. The purpose of this book is to present a comprehensive and up-to-date account of the major developments on optimal crossover designs.

The detailed discussion on optimal crossover designs begins in Chapter 2, where we introduce balanced and strongly balanced uniform crossover designs and present results on the optimality of these designs in some classes of competing designs. The optimality of nearly strongly balanced designs is also reviewed in Chapter 2. The model considered is the one given in (1.2.1), which is an additive, fixed effects model, with errors assumed to be uncorrelated random variables with zero mean and constant variance. This model includes only the first order carryover effects, apart from a general mean, the period effects, the subject effects and the direct effects of the treatments. Some methods of construction of these optimal designs are also described in Chapter 2.

The optimality results in Chapter 2 relate to crossover designs in which $p \geq t$, where p is the number of periods and t, the number of treatments. In practice, however, often an experimenter looks for designs for which the number of periods is small relative to the number of treatments as, in such an event, the experiment does not have to be continued for a long duration. Some optimality results on crossover designs satisfying $p < t$ are presented in Chapter 3 under the model (1.2.1). Construction of some optimal designs with $p < t$ are also given in Chapter 3.

An approach to arrive at optimal crossover designs via the approximate design theory was proposed by Kushner (1997a,b, 1998). This approach is quite general and is capable of producing optimal crossover designs in very general classes of competing designs. Results based on this approach have been reviewed in Chapter 4. The case of an arbitrary dispersion structure of the errors is taken up first by considering a positive definite matrix V as the dispersion matrix of the errors for every subject. Subsequently, simplified and readily usable versions of the results are given for the case of uncorrelated, homoscedastic errors, i.e., when $V = I_p$. These results relate to the problem of inference on direct effects. Similar results for inference on first order carryover effects are also presented in Chapter 4.

More complex additive models have been considered in Chapter 5 and optimal designs under such models have been discussed. Specifically, a model with self and mixed carryover effects is considered in Section 5.2 and optimal designs under such a model are identified. A model with carryover effects proportional to direct effects is considered in Section 5.3 and related optimal designs are reviewed.

The optimal crossover designs considered in Chapters 2–5 are all studied under models with no separate terms for interaction between treatments applied to the same subject in successive periods. While the assumption of absence of interaction may be justified in some situations, it may not be appropriate in all applications of crossover designs. Optimality aspects of crossover designs under non-additive models are considered in Chapter 6 where we revise the additive model (1.2.1) to include an interaction term between the treatment producing the direct effect and the treatment producing the carryover effect on a subject in a given period. Under this model, it is seen that some of the optimality results obtained under an additive model remain robust. We also consider in Chapter 6 a non-additive model in the presence of higher order, rather than only first order, carryovers and study optimal designs under such a model.

Chapter 7 summarizes some further developments on optimal crossover designs. The case of just two treatments is of considerable interest, especially in clinical trials. The optimality of two-treatment crossover designs, under correlated as well as uncorrelated errors, is taken up in Section 7.2. This material is related to the contents of Chapter 4. Section 7.3 deals with optimal designs under correlated errors and with more than two treatments. Optimal designs for test-control comparisons are discussed in Section 7.4. These designs play an important role in determining the relative performance of new test treatments vis-a-vis a standard treatment, called a control. Section 7.5 concerns crossover designs which remain efficient or optimal when the planned trial has to be terminated due to extraneous reasons. Such designs can be useful, for instance, in clinical trials if some of the subjects (patients) drop out before the planned completion of the experiment. Section 7.6 contains some additional comments.

Chapter 2

Optimality of Balanced and Strongly Balanced Designs

2.1 Introduction

An authoritative article by Hedayat and Afsarinejad (1975) stimulated interest in crossover designs, which were termed by these authors as repeated measurements designs. In particular, they defined various terms like uniformity and balance in this context, discussed some constructions and presented an extensive bibliography. Subsequently, Hedayat and Afsarinejad (1978) initiated the study of optimality of these designs and obtained optimal designs within a class of uniform designs with $p = t$. This condition of uniformity on the competing designs was relaxed later by Cheng and Wu (1980). They also defined the concept of strong balance and identified designs which are universally optimal in the global class $\Omega_{t,n,p}$. Other authors continued the study of balanced uniform designs and obtained more general optimality results. New classes of designs were also studied, e.g., the nearly strongly balanced designs by Kunert (1983) and the nearly balanced uniform designs by Bate and Jones (2006).

In Sections 2.3 and 2.4, we present results on the optimality properties of balanced and strongly balanced uniform crossover designs. Some results on nearly strongly balanced, nearly balanced and related designs are reviewed in Section 2.5. Finally, in Section 2.6, some construction procedures of the optimal designs considered in Sections 2.3 and 2.4 are presented. All the designs studied in this chapter have $p \geq t$. Designs for the case $p < t$ are considered in later chapters. The model considered is the traditional model (1.2.1) or its equivalent form (1.3.3). Throughout, we follow the notation introduced in Chapter 1.

2.2 Definitions and Some Basic Results

We begin with the following definitions.

Definition 2.2.1. A design $d \in \Omega_{t,n,p}$ is said to be uniform on periods if in each period, d assigns each treatment to the same number of subjects, i.e., $m_{dsi} = n/t,\ 1 \le s \le t,\ 1 \le i \le p$.

Definition 2.2.2. A design $d \in \Omega_{t,n,p}$ is said to be uniform on subjects if on each subject, d assigns each treatment to the same number of periods, i.e., $n_{dsj} = p/t,\ 1 \le s \le t,\ 1 \le j \le n$.

Definition 2.2.3. A design $d \in \Omega_{t,n,p}$ is said to be uniform if it is uniform on periods as well as on subjects.

Definition 2.2.4. A design $d \in \Omega_{t,n,p}$ is said to be balanced if in the order of application, no treatment precedes itself and each treatment is preceded by every other treatment a constant number of times, or equivalently, if among all the ordered pairs $\{d(i,j), d(i+1,j)\},\ 1 \le i \le p-1, 1 \le j \le n$, each ordered pair of distinct treatments occurs a constant number of times, say λ_1 times, i.e., $z_{dss'} = \lambda_1,\ z_{dss} = 0,\ \ 1 \le s, s' \le t;\ s \ne s'$.

Definition 2.2.5. A design $d \in \Omega_{t,n,p}$ is said to be strongly balanced if in the order of application, each treatment is preceded by every treatment (including itself) a constant number of times, or equivalently, if among all the ordered pairs $\{d(i,j), d(i+1,j)\},\ 1 \le i \le p-1, 1 \le j \le n$, each ordered pair of treatments (distinct or not) occurs a constant number of times, say λ_2 times, i.e., $z_{dss'} = \lambda_2,\ 1 \le s, s' \le t$.

It is clear from the above definitions that a necessary condition for a design $d \in \Omega_{t,n,p}$ to be uniform on periods is that $t|n$ and a necessary condition for a design to be uniform on subjects is that $t|p$, where for positive integers a, b, $a|b$ means that b is divisible by a. Thus, for a uniform design $d \in \Omega_{t,n,p}$,

$$n = \mu_1 t \text{ and } p = \mu_2 t \text{ for some integers } \mu_1, \mu_2 \ge 1.$$

Furthermore, a necessary condition for a design $d \in \Omega_{t,n,p}$ to be balanced is that $t(t-1)|n(p-1)$ and a necessary condition for d to be strongly balanced is that $t^2|n(p-1)$. It follows then that,

$$\lambda_1 = n(p-1)\{t(t-1)\}^{-1} = \mu_1(p-1)(t-1)^{-1} \text{ and}$$
$$\lambda_2 = n(p-1)t^{-2} = \mu_1(p-1)t^{-1}$$

are positive integers. Moreover, if a uniform design in $\Omega_{t,n,p}$ is also strongly balanced, then $p \ge 2t$.

Example 2.2.1. We give examples of some designs described in the above definitions. As in Chapter 1, the rows stand for the periods, columns represent the subjects and the treatments are denoted by 1,2,... etc.

$$d_1 \equiv \begin{matrix} 1 & 2 & 3 & 4 \\ 4 & 1 & 2 & 3 \\ 2 & 3 & 4 & 1 \\ 3 & 4 & 1 & 2 \end{matrix}, \qquad d_2 \equiv \begin{matrix} 1 & 1 & 1 & 2 & 2 & 2 & 3 & 3 & 3 \\ 1 & 2 & 3 & 1 & 2 & 3 & 1 & 2 & 3 \\ 2 & 2 & 2 & 3 & 3 & 3 & 1 & 1 & 1 \\ 2 & 3 & 1 & 2 & 3 & 1 & 2 & 3 & 1 \\ 3 & 3 & 3 & 1 & 1 & 1 & 2 & 2 & 2 \\ 3 & 1 & 2 & 3 & 1 & 2 & 3 & 1 & 2 \end{matrix},$$

$$d_3 \equiv \begin{matrix} 1 & 2 & 3 & 4 & 2 & 4 & 1 & 4 \\ 4 & 1 & 2 & 3 & 3 & 1 & 3 & 3 \\ 2 & 3 & 4 & 1 & 4 & 2 & 1 & 2 \\ 4 & 4 & 1 & 2 & 2 & 3 & 4 & 1 \end{matrix}, \qquad d_4 \equiv \begin{matrix} 1 & 2 & 3 & 3 & 1 & 2 \\ 2 & 3 & 1 & 2 & 3 & 1 \\ 3 & 1 & 2 & 1 & 2 & 3 \\ 3 & 1 & 2 & 1 & 2 & 3 \end{matrix}.$$

It is not hard to see that the design $d_1 \in \Omega_{4,4,4}$ is a balanced uniform design with $\mu_1 = \mu_2 = 1$ and $\lambda_1 = 1$ while $d_2 \in \Omega_{3,9,6}$ is a strongly balanced uniform design with $\mu_1 = 3, \mu_2 = 2$ and $\lambda_2 = 5$. Design $d_3 \in \Omega_{4,8,4}$ is *not* uniform on either subjects or periods, but is balanced with $\lambda_1 = 2$. Finally, $d_4 \in \Omega_{3,6,4}$ is a strongly balanced design with $\lambda_2 = 2$ and it is uniform on periods with $\mu_1 = 2$, but not uniform on subjects. Design d_4 is uniform on subjects in the first $3(= p - 1)$ periods and designs such as d_4 were termed extra-period designs by Patterson and Lucas (1959). See also Lucas (1957). □

The next Lemma follows from the definitions of the relevant designs given above.

Lemma 2.2.1. *Let d_1, d_2, d_3 and d_4 be crossover designs with t treatments, which are, respectively, uniform on periods, uniform, balanced uniform and strongly balanced uniform. Then*

$$d_1 \in \Omega_{t,n=\mu_1 t,p}, \qquad d_2 \in \Omega_{t,n=\mu_1 t,p=\mu_2 t},$$
$$d_3 \in \Omega_{t,n=\mu_1 t,p=\mu_2 t},\ (t-1)|\mu_1(p-1),\ d_4 \in \Omega_{t,n=\mu_1 t,p=\mu_2 t},\ t|\mu_1,\ \mu_2 \geq 2$$

and furthermore,

$$\begin{aligned} \boldsymbol{r}_{d_1} &= \boldsymbol{r}_{d_2} = \boldsymbol{r}_{d_3} = \boldsymbol{r}_{d_4} = \mu_1 p \mathbf{1}_t, \\ \bar{\boldsymbol{r}}_{d_1} &= \bar{\boldsymbol{r}}_{d_2} = \bar{\boldsymbol{r}}_{d_3} = \bar{r}_{d_4} = \mu_1(p-1)\mathbf{1}_t, \\ R_{d_1} &= R_{d_2} = R_{d_3} = R_{d_4} = \mu_1 p I_t, \\ \bar{R}_{d_1} &= \bar{R}_{d_2} = \bar{R}_{d_3} = \bar{R}_{d_4} = \mu_1(p-1)I_t, \\ M_{d_1} &= M_{d_2} = M_{d_3} = M_{d_4} = \mu_1 J_{tp}, \end{aligned}$$

$$\begin{aligned} \bar{M}_{d_1} &= \bar{M}_{d_2} = \bar{M}_{d_3} = \bar{M}_{d_4} = [\mathbf{0}_t \;\; \mu_1 J_{t,p-1}], \\ N_{d_2} &= N_{d_3} = N_{d_4} = \mu_2 J_{tn}, \\ Z_{d_3} &= \lambda_1 (J_t - I_t), \\ Z_{d_4} &= \lambda_2 J_t, \end{aligned} \tag{2.2.1}$$

where μ_1, μ_2, λ_1 *and* λ_2 *are positive integers.* □

Recalling that for a positive integer a, $H_a = I_a - a^{-1} J_a$, from (1.3.10), on applying (1.3.2) and (2.2.1) we have after simplification,

$$\begin{aligned} C_{d_2 11} &= \mu_1 p H_t \\ C_{d_2 12} &= Z_{d_2} - t^{-1} \mu_1 (p-1) J_t = C'_{d_2 21}, \\ C_{d_2 22} &= \mu_1 (p - 1 - p^{-1}) H_t, \\ C_{d_3 11} &= \mu_1 p H_t \\ C_{d_3 12} &= -\lambda_1 H_t \\ C_{d_3 22} &= \mu_1 (p - 1 - p^{-1}) H_t, \\ C_{d_4 11} &= \mu_1 p H_t, \\ C_{d_4 12} &= \mathbf{0}_{tt}, \\ C_{d_4 22} &= \mu_1 (p - 1 - p^{-1}) H_t. \end{aligned} \tag{2.2.2}$$

Then from (1.3.11) it is clear that, C_{d_3}, $\bar{C}_{d_3}$, C_{d_4}, $\bar{C}_{d_4}$ are all multiples of H_t and hence completely symmetric with forms as shown below. Moreover, since $C_{d_4 12} = \mathbf{0}$, the information matrices for d_4 are particularly simple.

$$\begin{aligned} C_{d_3} &= \mu_1 p \left[1 - (p-1)^2 (t-1)^{-2} (p^2 - p - 1)^{-1}\right] H_t, \\ \bar{C}_{d_3} &= \mu_1 \left[(p - 1 - p^{-1}) - (p-1)^2 (t-1)^{-2} p^{-1}\right] H_t, \\ C_{d_4} &= \mu_1 p H_t, \\ \bar{C}_{d_4} &= \mu_1 (p - 1 - p^{-1}) H_t. \end{aligned} \tag{2.2.3}$$

For any $d \in \Omega_{t,n,p}$, it is evident from (1.3.13) that both C_d and $\bar{C}_d$ are symmetric n.n.d. matrices. Furthermore, as seen in Chapter 1, these matrices have zero row sums. Hence they belong to the class $\mathcal{A}$ introduced in Section 1.4. As a result, in our search for optimal designs for direct and carryover effects over $\Omega_{t,n,p}$, we may use the universal optimality criterion and apply Theorem 1.4.1 to identify the optimal designs. The following results given by Cheng and Wu (1980) will be useful in the next section.

Lemma 2.2.2. *For any* $d \in \Omega_{t,n,p}$, *the following matrices are all n.n.d.*
(i) C_{d11}, *(ii)* C_{d22} , *(iii)* $R_d - n^{-1} M_d M'_d$, *(iv)* $R_d - p^{-1} N_d N'_d$,
(v) $\bar{R}_d - n^{-1} \bar{M}_d \bar{M}'_d$, *(vi)* $\bar{R}_d - p^{-1} \bar{N}_d \bar{N}'_d$, *(vii)* $M_d M'_d - p^{-1} M_d J_p M'_d$,
(viii) $N_d N'_d - n^{-1} N_d J_n N'_d$, *(ix)* $\bar{M}_d \bar{M}'_d - p^{-1} \bar{M}_d J_p \bar{M}'_d$,
(x) $\bar{N}_d \bar{N}'_d - n^{-1} \bar{N}_d J_n \bar{N}'_d$.

Proof. The assertions in (i) and (ii) are easy to see. To see the truth of (iii), observe that $R_d - n^{-1}M_dM_d'$ is the C-matrix for the direct effects when the carryover and subject effects are not in the model and thus, it is n.n.d. To prove (vii), note that for a positive integer a the matrix $I_a - a^{-1}J_a$ is n.n.d. Hence it follows that $M_dM_d' - p^{-1}M_dJ_pM_d' = M_d(I_p - p^{-1}J_p)M_d'$ is n.n.d. The other results in the lemma follow similarly. □

Lemma 2.2.3. *For any positive integers a, b and nonnegative integers $u_1, \ldots, u_a$, the minimum of $u_1^2 + \cdots + u_a^2$, subject to $u_1 + \cdots + u_a = b$, is obtained when $b - a[b/a]$ of the u_i's are each equal to $[b/a] + 1$ and the rest are each equal to $[b/a]$, where $[z]$ denotes the largest integer not exceeding $z (> 0)$.* □

The above result is well known; a proof can be found, e.g., in Cheng (1978).

Lemma 2.2.4. *Let $d^* \in \Omega_{t,n,p}$ be a uniform design. Then d^* maximizes $\mathrm{tr}(C_{d11})$ and $\mathrm{tr}(C_{d22})$ over $\Omega_{t,n,p}$.*

Proof. We present the proof for C_{d22}; the proof for C_{d11} is similar. From (1.3.10), for any $d \in \Omega_{t,n,p}$,

$$\begin{aligned}\mathrm{tr}(C_{d22}) &= n(p-1) - n^{-1}\sum_{s=1}^{t}\sum_{i=1}^{p-1} m_{dsi}^2 - p^{-1}\sum_{s=1}^{t}\sum_{j=1}^{n}\bar{n}_{dsj}^2 + (np)^{-1}\sum_{s=1}^{t}\bar{r}_{ds}^2 \\ &= n(p-1) - p^{-1}\sum_{s=1}^{t}\sum_{j=1}^{n}\bar{n}_{dsj}^2 - n^{-1}\left[\sum_{s=1}^{t}\sum_{i=1}^{p-1}(m_{dsi} - (p-1)^{-1}\bar{r}_{ds})^2\right. \\ &\quad \left. + \{(p-1)^{-1} - p^{-1}\}\sum_{s=1}^{t}\bar{r}_{ds}^2\right],\end{aligned}$$

as $\bar{r}_{ds} = \sum_{i=1}^{p-1} m_{dsi}$. Now, from Lemma 2.2.1, for the uniform design d^*, $m_{d^*si} = (n/t)$ for all s, i and $\bar{r}_{d^*s} = n(p-1)/t$ for all s, $1 \le s \le t, 1 \le i \le p$. Therefore, $\sum_{s=1}^{t}\sum_{i=1}^{p-1}\left(m_{d^*si} - (p-1)^{-1}\bar{r}_{d^*s}\right)^2 = 0$, and thus for proving the result, it is enough to show that d^* minimizes $\sum_{s=1}^{t}\sum_{j=1}^{n}\bar{n}_{dsj}^2$ and $\sum_{s=1}^{t}\bar{r}_{ds}^2$. This is clear from Lemma 2.2.3 since

$$\sum_{s=1}^{t}\sum_{j=1}^{n}\bar{n}_{dsj} = n(p-1) = \sum_{s=1}^{t}\bar{r}_{ds},$$

$|\bar{n}_{d^*sj} - \bar{n}_{d^*s'j'}| \le 1$ for all $(s,j) \ne (s',j')$ and $\bar{r}_{d^*s}$ are equal for all s. □

Lemma 2.2.5. *Let $d^* \in \Omega_{t,n,p}$ be a design which is uniform on periods and uniform on subjects in the first $p-1$ periods. Then d^* maximizes $\mathrm{tr}(C_{d11})$ and $\mathrm{tr}(C_{d22})$ over $\Omega_{t,n,p}$.*

Proof. In this case we observe that $\bar{n}_{d^*sj}$ is a constant for all s, j, while $|n_{d^*sj} - n_{d^*s'j'}| \leq 1$ for all $(s, j) \neq (s', j')$. Now, the proof follows as in Lemma 2.2.4. □

We only state the following results on constrained minimization and refer to Cheng and Wu (1980) for their proofs.

Lemma 2.2.6. *For reals $a_1, \ldots, a_t$, the minimum of $\sum_{s=1}^t a_s^2$, subject to $\sum_{s=1}^t a_s = b$, is attained when $a_s = b/t$, $1 \leq s \leq t$, and this minimum is equal to b^2/t.* □

Lemma 2.2.7. *For reals $a_1, \ldots, a_t$, the minimum of $\sum_{s=1}^t a_s^2/r_s$, $r_s > 0$, subject to $\sum_{s=1}^t a_s = b$ and $\sum_{s=1}^t r_s = c > 0$, is attained when $a_s = br_s/c$, $1 \leq s \leq t$, and this minimum is equal to b^2/c.* □

Lemma 2.2.8. *Suppose the sequences $\{a_j\}$ and $\{b_j\}$, $1 \leq j \leq n$, of reals are such that $(a_j - a_{j'})(b_j - b_{j'}) \geq 0$, for all $1 \leq j, j' \leq n$. Then $\sum_{j=1}^n a_j b_j \geq n^{-1}(\sum_{j=1}^n a_j)(\sum_{j=1}^n b_j)$.* □

Lemma 2.2.9. *For $0 \leq c \leq [2t(p-1)\mu_1^2]^{-1}$ and reals $\{a_{si}\}$, the minimum of*

$$\sum_{s=1}^{t}\sum_{i=1}^{p-1} a_{si}^2 + c\left(\sum_{s=1}^{t}\sum_{i=2}^{p} a_{si}a_{s,i-1}\right)^2,$$

subject to $\sum_{s=1}^t a_{si} = \mu_1 t$ for $1 \leq i \leq p$, is attained when $a_{si} = \mu_1$ for all $1 \leq s \leq t$, $1 \leq i \leq p$. □

Lemma 2.2.10. *For $0 \leq c \leq \left[tp(p-1)\mu_1^2\right]^{-1}$ and reals $\{a_{si}\}$, the minimum of*

$$\sum_{s=1}^{t}\sum_{i=1}^{p-1} a_{si}^2 - p^{-1}\sum_{s=1}^{t}\left(\sum_{i=1}^{p-1} a_{si}\right)^2 + c\left(\sum_{s=1}^{t}\sum_{i=2}^{p} a_{si}a_{s,i-1}\right)^2,$$

subject to $\sum_{s=1}^t a_{si} = \mu_1 t$ for $1 \leq i \leq p$, is attained when $a_{si} = \mu_1$ for all $1 \leq s \leq t$ and all $1 \leq i \leq p$. □

From the next section onwards, we review various available results on the optimality of crossover designs. A main hurdle in obtaining such results in the class of all designs $\Omega_{t,n,p}$ is that the information matrix for direct effects, C_d, involves the g-inverse of C_{d22} ; see (1.3.11). Since C_{d22} is design dependent, finding a general expression for its g-inverse for an arbitrary design in this class can pose a serious challenge. This makes the determination of $\mathrm{tr}(C_d)$ for an arbitrary design d difficult. The problem has been tackled in the literature in different ways, giving rise to different optimality results.

For instance, some authors have restricted the class of competing designs to certain suitable subclasses of $\Omega_{t,n,p}$, while others have considered C_d under a simpler model than the one under study and then verified appropriate orthogonality conditions as in Lemma 1.4.1. Some authors have also used Lemma 1.4.2, or a generalization of it, to obtain a simple upper bound for $\operatorname{tr}(C_d)$ for arbitrary d and this helped in identifying designs satisfying the conditions of Theorem 1.4.1. Many of the results that follow in this and the subsequent chapters hinge on these approaches.

2.3 Optimality of Balanced Uniform Designs

In this section, results on the optimality of balanced uniform designs are presented following Hedayat and Afsarinejad (1978), Cheng and Wu (1980), Kunert (1984b) and Hedayat and Yang (2003, 2004).

As remarked in the preceding section, for an arbitrary $d \in \Omega_{t,n,p}$, g-inverses of C_{d11} and C_{d22} are intractable and so obtaining C_d and $\bar{C}_d$ can pose a problem. Hedayat and Afsarinejad (1978) overcame this difficulty by considering a subclass of uniform designs in $\Omega_{t,n=\mu_1 t,p=t}$. This led to considerable simplification since for all designs in this subclass, both C_d and $\bar{C}_d$ have simple expressions, thereby facilitating the application of Theorem 1.4.1 in proving optimality. As we shall see later in this section, the result of Hedayat and Afsarinejad (1978) for direct effects was subsequently generalized; we present their proof in order to highlight its simplicity and directness. They considered the subclass of designs

$$\Omega^*_{t,n,t} = \{d \,:\, d \in \Omega_{t,n,p},\ n = \mu_1 t,\ p = t,\ d \text{ is uniform}\},$$

and obtained the following result.

Theorem 2.3.1 *Let $d^* \in \Omega^*_{t,\mu_1 t,t}$ be a balanced design. Then d^* is universally optimal for the estimation of direct effects and also for the estimation of carryover effects over $\Omega^*_{t,\mu_1 t,t}$.*

Proof. It can be verified using Lemma 2.2.1 and (1.3.11) that for any design $d \in \Omega^*_{t,\mu_1 t,t}$

$$C_d = nI_t - t\mu_1^{-1}(t^2-t-1)^{-1}Z_d Z_d' + \mu_1(2-t)(t^2-t-1)^{-1}J_t,$$

$$\bar{C}_d = t^{-1}\mu_1(t^2-t-1)I_t - n^{-1}Z_d Z_d' + t^{-2}\mu_1(2-t)J_t,$$

and as usual, $C_d\mathbf{1}_t = \mathbf{0}_t = \bar{C}_d\mathbf{1}_t$, for all $d \in \Omega^*_{t,\mu_1 t,t}$. Again, from Lemma 2.2.1, since d^* is balanced, $Z_{d^*} = \lambda_1(J_t - I_t)$ and so, both C_{d^*} and $\bar{C}_{d^*}$ are completely symmetric.

From the form of C_d above, it is clear that in order to maximize $\operatorname{tr}(C_d)$, we need to minimize $\operatorname{tr}(Z_d Z_d')$. Now with $p = t$, it follows that $\lambda_1 = \mu_1$ and it can be seen that $Z_d \mathbf{1}_t = \mu_1(t-1)\mathbf{1}_t$ for every $d \in \Omega^*_{t,\mu_1 t,t}$, invoking the uniformity of any such design. This in turn, implies that $\operatorname{tr}(Z_d Z_d')$ is minimized if $Z_d = \mu_1(J_t - I_t) = Z_{d^*}$. Thus,

$$\operatorname{tr}(C_{d^*}) = \max_{d\in\Omega^*} \operatorname{tr}(C_d), \text{ and similarly, } \operatorname{tr}(\bar{C}_{d^*}) = \max_{d\in\Omega^*} \operatorname{tr}(\bar{C}_d).$$

In view of Theorem 1.4.1, the proof is now complete. □

Cheng and Wu (1980) extended Theorem 2.3.1 by relaxing the condition of uniformity on the competing designs. However, some restrictions were still needed and they considered the subclass of designs as defined below:

$$\Lambda_{t,n,p} = \{d \,:\, d \in \Omega_{t,n,p},\ z_{dss} = 0,\ 1 \le s \le t\},$$

i.e., $\Lambda_{t,n,p}$ consists of designs in which no treatment is assigned to two consecutive periods on the same subject. They proved the optimality of balanced uniform designs over certain subclasses of $\Lambda_{t,n,p}$. These results are given below in Theorems 2.3.2–2.3.4. The first two theorems pertain to optimality for carryover effects and for proving these, they used an interesting technique which allowed them to avoid the computation of a g-inverse of C_{d11} for arbitrary d and work with only the inverse of a diagonal matrix.

Theorem 2.3.2. *Let $d^* \in \Lambda_{t,n=\mu_1 t,p=\mu_2 t}$ be a balanced uniform design, where $t \ge 3$. Then d^* is universally optimal for the estimation of carryover effects over the subclass of designs $\Lambda_0 \subset \Lambda_{t,\mu_1 t,\mu_2 t}$, where Λ_0 consists of designs in which each treatment is equally replicated in the first $p-1$ periods.*

Proof. By definition, $\Lambda_0 = \{d : d \in \Lambda_{t,\mu_1 t,\mu_2 t};\ \bar{\boldsymbol{r}}_d = \bar{r}\mathbf{1}_t \text{ for some } \bar{r}\}$. Clearly, $\bar{r} = \mu_1(p-1)$. Now, from (2.2.3), $\bar{C}_{d^*}$ is completely symmetric. So, by Theorem 1.4.1, to prove the theorem, we need to show that d^* maximizes $\operatorname{tr}(\bar{C}_d)$ over Λ_0.

The fact that $M_d M_d'$ is n.n.d, in conjunction with Lemma 2.2.2(viii) and (1.3.10), shows that

$$C_{d11} \le R_d, \text{ for all } d \in \Omega_{t,n,p}.$$

Hence there always exists a g-inverse C_{d11}^- such that $C_{d11}^- \ge R_d^{-1}$ for all $d \in \Omega_{t,n,p}$ (*cf.* Wu (1980)). Recalling from Lemma 2.2.1 and (2.2.2) that $C_{d^*11}^- = R_{d^*}^{-1}$, in order to show that d^* maximizes $\operatorname{tr}(\bar{C}_d)$ over Λ_0 it is now enough to show that d^* maximizes $\operatorname{tr}(C_{d22} - C_{d21} R_d^{-1} C_{d12})$ over Λ_0. This latter expression is simpler as it does not involve any g-inverse and so its maximization is easier. In the rest of the proof we discuss this maximization problem.

From (1.3.10), it follows that for every $d \in \Lambda_0$,

$$\operatorname{tr}(C_{d22}) = \sum_{s=1}^{t} \bar{r}_{ds} - n^{-1} \sum_{s=1}^{t} \sum_{i=1}^{p-1} m_{dsi}^2 - p^{-1} \sum_{s=1}^{t} \sum_{j=1}^{n} \bar{n}_{dsj}^2 + (np)^{-1} \sum_{s=1}^{t} \bar{r}_{ds}^2 \tag{2.3.1}$$

and

$$\begin{aligned} \operatorname{tr}(C_{d21} R_d^{-1} C_{d12}) &= \sum_{s=1}^{t} r_{ds}^{-1} \sum_{s'=1}^{t} \Bigg(z_{dss'} - n^{-1} \sum_{i=2}^{p} m_{dsi} m_{ds',i-1} \\ &\quad - p^{-1} \sum_{j=1}^{n} n_{dsj} \bar{n}_{ds'j} + (np)^{-1} r_{ds} \bar{r}_{ds'} \Bigg)^2 \qquad (2.3.2) \\ &= \sum_{s=1}^{t} r_{ds}^{-1} \sum_{s'=1}^{t} (z_{dss'} - q_{dss'})^2, \end{aligned}$$

where

$$q_{dss'} = n^{-1} \sum_{i=2}^{p} m_{dsi} m_{ds',i-1} + p^{-1} \sum_{j=1}^{n} n_{dsj} \bar{n}_{ds'j} - (np)^{-1} r_{ds} \bar{r}_{ds'}. \tag{2.3.3}$$

Also for any $d \in \Lambda_0$, the row and column sums of C_{d12} are zero, so that

$$\text{(a)} \quad \sum_{s=1}^{t} (z_{dss'} - q_{dss'}) = \sum_{s'=1}^{t} (z_{dss'} - q_{dss'}) = 0.$$

In addition, every $d \in \Lambda_0$ satisfies the following conditions:

$$\text{(b)} \ z_{dss} = 0, \ 1 \le s \le t,$$

$$\text{(c)} \ \sum_{s=1}^{t} m_{dsi} = n = \mu_1 t, \ 1 \le i \le p,$$

$$\text{(d)} \ \sum_{s=1}^{t} r_{ds} = np = n\mu_2 t.$$

Now, since $\sum_{j=1}^{n} \bar{n}_{dsj} = \bar{r}_{ds}$ is a constant for all $d \in \Lambda_0$, $1 \le s \le t$, and $|\bar{n}_{d^*sj} - \bar{n}_{d^*s'j'}| \le 1$ for all $(s, j) \ne (s', j')$, by Lemma 2.2.3, $\sum_{s=1}^{t} \sum_{j=1}^{n} \bar{n}_{dsj}^2$ is minimized by d^* over Λ_0. Hence, from (2.3.1) and (2.3.2), it is enough to show that d^* minimizes

$$n^{-1} \sum_{s=1}^{t} \sum_{i=1}^{p-1} m_{dsi}^2 + \sum_{s=1}^{t} r_{ds}^{-1} \sum_{s'=1}^{t} (z_{dss'} - q_{dss'})^2, \tag{2.3.4}$$

subject to (a)–(d).

Cheng and Wu (1980) showed that d^* minimizes the expression in (2.3.4); the necessary steps are indicated below.

Step 1. By (b), (a) reduces to

$$(\text{a}') \quad \sum_{s'=1, s'\neq s}^{t} (z_{dss'} - q_{dss'}) = q_{dss}.$$

Applying Lemma 2.2.6, the minimum of

$$\sum_{s'=1}^{t} (z_{dss'} - q_{dss'})^2 = \sum_{s'=1, s'\neq s}^{t} (z_{dss'} - q_{dss'})^2 + q_{dss}^2$$

subject to (a$'$), is attained when $z_{dss'} - q_{dss'} = q_{dss}/(t-1)$ for all $s' \neq s$ and the minimum value is $q_{dss}^2/(t-1) + q_{dss}^2 = q_{dss}^2 t(t-1)^{-1}$. Since d^* is balanced and also uniform, it follows from Lemma 2.2.1 and (2.3.3) that for $d \equiv d^*$, $z_{d^*ss'} - q_{d^*ss'} = q_{d^*ss}/(t-1)$, $(1 \leq s, s' \leq t; s \neq s')$ and so, this minimum is attained.

Step 2. Applying Lemma 2.2.7 with $c = np = n\mu_2 t$, it can be seen that the minimum of

$$\sum_{s=1}^{t} r_{ds}^{-1} q_{dss}^2$$

subject to (d) is equal to $(n\mu_2 t)^{-1} \left(\sum_{s=1}^{t} q_{dss}\right)^2$. Again, for d^*, we have $r_{d^*s} = n\mu_2$ and $q_{d^*ss} = t^{-1} \sum_{s=1}^{t} q_{d^*ss}$, and so this minimum is attained.

Step 3. From (2.3.3), since $\sum_{j=1}^{n} n_{dsj} = r_{ds}$ and $\sum_{j=1}^{n} \bar{n}_{dsj} = \bar{r}_{ds}$, it follows that

$$\begin{aligned} \sum_{s=1}^{t} q_{dss} = {} & n^{-1} \sum_{s=1}^{t} \sum_{i=2}^{p} m_{dsi} m_{ds,i-1} \\ & + p^{-1} \sum_{s=1}^{t} \sum_{j=1}^{n} (n_{dsj} - n^{-1} r_{ds})(\bar{n}_{dsj} - n^{-1} \bar{r}_{ds}). \end{aligned} \tag{2.3.5}$$

Now, for any j, j', if $n_{dsj} > n_{dsj'}$, then $n_{dsj} \geq n_{dsj'} + 1$ and so, $\bar{n}_{dsj} \geq \bar{n}_{dsj'}$, implying that the sequences $\{n_{dsj}\}$ and $\{\bar{n}_{dsj}\}$, $1 \leq j \leq n$, are such that $(n_{dsj} - n_{dsj'})(\bar{n}_{dsj} - \bar{n}_{dsj'}) \geq 0$ for all $j, j', 1 \leq j, j' \leq n$. Similarly for the other cases. Hence, by Lemma 2.2.8,

$$\sum_{j=1}^{n} (n_{dsj} - n^{-1} r_{ds})(\bar{n}_{dsj} - n^{-1} \bar{r}_{ds}) \geq 0. \tag{2.3.6}$$

From Lemma 2.2.1, it can be seen that for the design d^*, $n_{d^*sj} = \mu_2, r_{d^*s} = n\mu_2$ and so, equality holds in (2.3.6) when $d \equiv d^*$. Therefore, from (2.3.5),

$$\left(\sum_{s=1}^{t} q_{dss}\right)^2 \geq \left(n^{-1}\sum_{s=1}^{t}\sum_{i=2}^{p} m_{dsi}m_{ds,i-1}\right)^2,$$

with equality when $d \equiv d^*$.

After Steps 1, 2 and 3, it follows from (2.3.4) that now it is enough to show that d^* minimizes

$$n^{-1}\sum_{s=1}^{t}\sum_{i=1}^{p-1} m_{dsi}^2 + \left\{t(t-1)^{-1}(n\mu_2 t)^{-1}\left(n^{-1}\sum_{s=1}^{t}\sum_{i=2}^{p} m_{dsi}m_{ds,i-1}\right)^2\right\}$$

$$= n^{-1}\left[\sum_{s=1}^{t}\sum_{i=1}^{p-1} m_{dsi}^2 + \{(t-1)\mu_2 n^2\}^{-1}\left\{\sum_{s=1}^{t}\sum_{i=2}^{p} m_{dsi}m_{ds,i-1}\right\}^2\right], \tag{2.3.7}$$

subject to the constraint (c).

Step 4. Note that $\{(t-1)\mu_2 n^2\}^{-1} \leq \{2t(p-1)\mu_1^2\}^{-1}$ for all $t \geq 3$. Then by Lemma 2.2.9, the minimum of (2.3.7) subject to (c) is attained when $m_{dsi} = \mu_1$ for all s, i. Since $m_{d^*si} = \mu_1$ for all s, i, this minimum is attained by d^*. This completes the proof. □

Remark 2.3.1. When $p = t$, Cheng and Wu (1980) showed that the restriction to the subclass Λ_0 as required in Theorem 2.3.2, may be avoided and the result is valid over the larger class $\Lambda_{t,n=\mu_1 t,p=t}$. This is shown in the next theorem. For optimal estimation of carryover effects over a still wider class, see Theorem 2.3.12 and the comments preceding it. □

Theorem 2.3.3. *Let d^* be a balanced uniform design in $\Lambda_{t,\mu_1 t,t}$, $t \geq 3$. Then d^* is universally optimal for the estimation of carryover effects over $\Lambda_{t,\mu_1 t,t}$.*

Proof. As in Theorem 2.3.2, $\bar{C}_{d^*}$ is completely symmetric but here we need to show that d^* maximizes $\text{tr}(C_{d22} - C_{d21}R_d^{-1}C_{d12})$ over $\Lambda_{t,\mu_1 t,t}$. The essential difference between the proof of this theorem and that of Theorem 2.3.2 is that here we do not have the constancy of $\bar{r}_{ds}$ for all s, though $\sum_{s=1}^{t}\bar{r}_{ds}$ is a constant as usual. So, instead of the expression in (2.3.4), from (2.3.1) and (2.3.2) it is easy to see that now we need to minimize

$$n^{-1}\sum_{s=1}^{t}\sum_{i=1}^{p-1} m_{dsi}^2 + \sum_{s=1}^{t} r_{ds}^{-1}\sum_{s'=1}^{t}(z_{dss'} - q_{dss'})^2 - (np)^{-1}\sum_{s=1}^{t}\bar{r}_{ds}^2,$$

subject to the same conditions (a)–(d). Proceeding as in Steps 1–3 of the proof of Theorem 2.3.2, and writing $\bar{r}_{ds} = \sum_{i=1}^{p-1} m_{dsi}$, it is clear that it is enough to show that d^* minimizes

$$n^{-1}\left[\sum_{s=1}^{t}\sum_{i=1}^{p-1} m_{dsi}^2 + \{(t-1)\mu_2 n^2\}^{-1}\left\{\sum_{s=1}^{t}\sum_{i=2}^{p} m_{dsi}m_{ds,i-1}\right\}^2 \right. \\ \left. -p^{-1}\sum_{s=1}^{t}\left(\sum_{i=1}^{p-1} m_{dsi}\right)^2\right], \tag{2.3.8}$$

subject to the constraint (c).

Now, $\{(t-1)\mu_2 n^2\}^{-1} \le \{tp(p-1)\mu_1^2\}^{-1}$ when $p = t$, i.e., $\mu_2 = 1$. Hence, by Lemma 2.2.10, the minimum of (2.3.8) subject to (c) is attained when $m_{dsi} = \mu_1$ for all s, i. Noting that $m_{d^*si} = \mu_1$ for all s, i, the proof is completed. □

Remark 2.3.2. The proofs of Theorems 2.3.2 and 2.3.3 made use of the fact that $C_{d^*11}^{-} - R_{d^*}^{-1} = \mathbf{0}$ and this simplified the proofs considerably. However, a similar simplification cannot be achieved while attempting to prove the optimality of d^* for direct effects, since from (2.2.2) and Lemma 2.2.1,

$$C_{d^*22}^{-} - \bar{R}_{d^*}^{-1} = \mu_1^{-1}\left[(p-1-p^{-1})^{-1}I_t - (p-1)^{-1}I_t\right] \neq \mathbf{0}.$$

Consequently, the minimization of $\operatorname{tr}(C_{d12}C_{d22}^{-}C_{d21})$ over $\Lambda_{t,\mu_1 t,\mu_2 t}$ is not equivalent to the minimization of $\operatorname{tr}(C_{d12}\bar{R}_d^{-1}C_{d21})$. So, in order to prove the optimality results for direct effects, Cheng and Wu (1980) restricted the class of competing designs to:

$$\Lambda_{t,\mu_1 t,\mu_2 t}^{*} = \{d : d \in \Lambda_{t,\mu_1 t,\mu_2 t},\ d \text{ is uniform on subjects and uniform on the last period}\}.$$

We state the result and refer to Cheng and Wu (1980) (and its corrigendum (1983)) for the proof. □

Theorem 2.3.4. *Let $d^* \in \Lambda_{t,\mu_1 t,\mu_2 t}$ be a balanced uniform design. Then d^* is universally optimal for the estimation of direct effects over $\Lambda_{t,\mu_1 t,\mu_2 t}^{*}$.* □

All the optimality results of the preceding theorems are valid over specific subclasses of $\Omega_{t,n,p}$. Kunert (1984b) studied the case $p = t$, and established the universal optimality of balanced uniform designs for the estimation of direct effects over the entire class of competing designs $\Omega_{t,n=\mu_1 t,p=t}$ when $\mu_1 = 1$ or 2.

Theorems 2.3.5–2.3.8 and 2.3.12 below are due to Kunert (1984b). The proofs of the first two rest on the following lemmas. For proofs of these lemmas, we refer to Kunert (1984b).

Lemma 2.3.1. *Let* $p = t > 2$ *and* $t|n$. *For any* $d \in \Omega_{t,n,t}$, *define* $x_{dsj} = n_{dsj} - 1$ *for every pair* (s, j), $1 \leq s \leq t$, $1 \leq j \leq n$. *Assume that the sum of all positive* x_{dsj} *is equal to an integer* b. *Let* $d^* \in \Omega_{t,n,t}$ *be a balanced uniform design. Then*
(i) a necessary condition for $\mathrm{tr}(C_d) \geq \mathrm{tr}(C_{d^*})$ *is that* $b < n/2 + n/(2t)$, *and*
(ii) for $1 \leq b \leq t$, $\mathrm{tr}(C_d) \leq n(t-1) - n/t - b(t-1)/\{t(n+1)\}$. □

Lemma 2.3.2. *Let* $d \in \Omega_{t,n,t}$ *be uniform on subjects but not on periods. Then the sum of all positive terms* $(m_{dsi} - n/t)$ *equals* q *where* $q \geq 2$ *and* $\mathrm{tr}(C_d) \leq n(t-1) - n/t - q\{2t(t-1) - 4n\}/\{nt(t-1)\}$. □

The next two theorems give results for direct effects when $n = t$ and $n = 2t$, respectively.

Theorem 2.3.5. *Let* $t = n = p > 2$ *and let* $d^* \in \Omega_{t,t,t}$ *be a balanced uniform design. Then* d^* *is universally optimal for the estimation of direct effects over* $\Omega_{t,t,t}$.

Proof. Since d^* does not exist in $\Omega_{3,3,3}$ (see Section 2.6), we assume that $t > 3$. Suppose there is a design $d_0 \in \Omega_{t,t,t}$ such that $\mathrm{tr}(C_{d_0}) > \mathrm{tr}(C_{d^*})$. Then by Theorem 2.3.4, it follows that d_0 must be either (a) not uniform on subjects or, (b) uniform on subjects but not uniform on periods.

If d_0 satisfies (b), then from Lemma 2.3.2 we have

$$\begin{aligned}\mathrm{tr}(C_{d_0}) &\leq t(t-1) - 1 - q\{2t(t-1) - 4t\}/\{t^2(t-1)\}\\ &= t(t-1) - 1 - 2q(t-3)/\{t(t-1)\},\\ &< \mathrm{tr}(C_{d^*})\end{aligned}$$

where the last inequality follows on noting that for $d^* \in \Omega_{t,t,t}$, from (2.2.3),

$$\mathrm{tr}(C_{d^*}) = t(t-1) - (t-1)/(t-1-t^{-1}) = t(t-1) - 1 - (t^2 - t - 1)^{-1}$$

and, for $t > 3$, $2q(t-3)/\{t(t-1)\} > (t^2 - t - 1)^{-1}$ since $q > 1$. Similarly, if d_0 satisfies (a) then by Lemma 2.3.1, one can check that $\mathrm{tr}(C_{d_0}) < \mathrm{tr}(C_{d^*})$.

Again, (2.2.3) shows that C_{d^*} is completely symmetric. The result now follows by Theorem 1.4.1. □

Example 2.3.1. Let $t = n = p = 4$. The optimality of the design d_1 of Example 2.2.1 follows from Theorem 2.3.5. □

When $n = 2t$, Kunert (1984b) showed that optimality can be proved over the entire class for $t \geq 6$. This result is stated below.

Theorem 2.3.6. *Let $n = 2t$, $p = t$, $t \geq 6$ and let $d^* \in \Omega_{t,2t,t}$ be a balanced uniform design. Then d^* is universally optimal for the estimation of direct effects over $\Omega_{t,2t,t}$.* □

The proof of Theorem 2.3.6 for $t \geq 7$ is analogous to that of Theorem 2.3.5. For $t = 6$, a modification is needed; for details, see Kunert (1984b).

Remark 2.3.3. The design d^* of Theorem 2.3.6 is also optimal for $t = 5$ as shown later in Theorem 2.3.9. If n is large, one can find designs which are better than d^*. Towards this, we first give a definition and a lemma. □

Definition 2.3.1. Let $n = t(t-1)$ and $d^* \in \Omega_{t,n,t}$ be a balanced uniform design such that every ordered pair of distinct treatments appears once over the last two periods. Let $f \in \Omega_{t,n,t}$ be a design consisting of the first $(t-1)$ periods of d^* and a last period equal to the $(t-1)$th period of d^*. Then f is called an orthogonal residual effects design.

Example 2.3.2. The following design f is an example of an orthogonal residual effects design with $t = p = 3$ and $n = 6$ obtained from the balanced uniform design d^* shown alongside.

$$d^* = \begin{matrix} 1 & 2 & 3 & 1 & 2 & 3 \\ 2 & 3 & 1 & 3 & 1 & 2 \\ 3 & 1 & 2 & 2 & 3 & 1 \end{matrix}, \quad f \equiv \begin{matrix} 1 & 2 & 3 & 1 & 2 & 3 \\ 2 & 3 & 1 & 3 & 1 & 2 \\ 2 & 3 & 1 & 3 & 1 & 2 \end{matrix}.$$

□

Theorem 2.3.7. *Let $t = p > 2$ and $n = \mu_1 t$ with $n > \{t(t-1)\}^2/2$. Let a design $g \in \Omega_{t,n,t}$ be such that*
(i) the first $t(t-1)$ subjects of g form an orthogonal residual effects design $f \in \Omega_{t,t(t-1),t}$, and
(ii) the remaining subjects of g form a balanced uniform design in $\Omega_{t,n-t(t-1),t}$.
Then for the estimation of direct effects, no balanced uniform design in $\Omega_{t,n,t}$ is universally optimal and any such design is dominated by g. □

Remark 2.3.4. By Theorem 2.3.7, a balanced uniform design d^* is no longer universally optimal over $\Omega_{t,n=\mu_1 t,t}$ if n is sufficiently large. However, the following theorem shows that d^* is still highly efficient for all $n(= \mu_1 t)$. □

Theorem 2.3.8. *If a balanced uniform design $d^* \in \Omega_{t,n=\mu_1 t,t}$ exists, then*

$$\frac{\mathrm{tr}(C_{d^*})}{\sup\limits_{d \in \Omega_{t,n=\mu_1 t,t}} [\mathrm{tr}(C_d)]} \geq 1 - \frac{1}{(t^2 - t - 1)^2}$$

for every $n = \mu_1 t$, where μ_1 is a positive integer. □

Remark 2.3.5. Theorem 2.3.7 gives a lower bound to n such that a balanced uniform design d^* fails to be optimal if n crosses this bound. For $p = t$, Kunert (1984b) gave an upper bound to n $(= t(t-1)/2)$ such that for all t and n satisfying this bound, the design d^* is universally optimal for direct effects in a certain subclass of $\Omega_{t,n,t}$. Hedayat and Yang (2003) extended this result by removing the restriction on the competing class of designs and showed that d^* remains optimal over the entire class $\Omega_{t,n,t}$ as long as n is within this bound. Their result is given below as Theorem 2.3.9. This result thus also extends Theorem 2.3.6 to all $t \geq 5$ and Theorem 2.3.5 follows from it. We sketch the proof as in Hedayat and Yang (2003). This theorem can also be deduced from Theorem 3.2.5 in Chapter 3 on noting that in our present context with $p = t$, the Stufken designs considered there reduce to balanced uniform designs if and only if $n \leq t(t-1)/2$. □

Theorem 2.3.9. *Let $t = p > 2, n = \mu_1 t$ with $n \leq t(t-1)/2$. Then a balanced uniform design $d^* \in \Omega_{t,n,t}$ is universally optimal for the direct effects over $\Omega_{t,n,t}$.*

Proof. From (2.2.3), C_{d^*} is completely symmetric. Therefore, to complete the proof, it suffices to show that $\operatorname{tr}(C_d) \leq \operatorname{tr}(C_{d^*})$ for all $d \in \Omega_{t,\mu_1 t,t}$.

From (2.2.3), for $p = t$, $\operatorname{tr}(C_{d^*}) = n(t-1) - \dfrac{n(t-1)}{t^2-t-1}$. Now, with b as defined in Lemma 2.3.1, we have that if $b \geq n$, then $\operatorname{tr}(C_d) \leq \operatorname{tr}(C_{d^*})$. So, to complete the proof it suffices to show that for $b < n$,

$$\operatorname{tr}(C_d) \leq n(t-1) - \frac{n(t-1)}{t^2-t-1}.$$

Using Lemma 1.4.2, one can show that for any $d \in \Omega_{t,\mu_1 t,t}$,

$$\operatorname{tr}(C_d) \leq q_{11}(d) - \frac{q_{12}^2(d)}{q_{22}(d)}, \tag{2.3.9}$$

where

$$q_{11}(d) = \mu_1 t^2 - t^{-1} \sum_{j=1}^{n} \sum_{s=1}^{t} n_{dsj}^2$$

$$q_{12}(d) = \sum_{s=1}^{t} z_{dss} - t^{-1} \sum_{j=1}^{n} \sum_{s=1}^{t} n_{dsj} \tilde{n}_{dsj}$$

$$q_{22}(d) = \mu_1 t(t-1)(1-t^{-2}) - t^{-1} \sum_{j=1}^{n} \sum_{s=1}^{t} \tilde{n}_{dsj}^2.$$

Hedayat and Yang (2003) have shown that for $b \in [0, n)$,

$$q_{11}(d) \leq n(t-1) - \frac{2b}{t},$$
$$q_{12}^2(d) \geq \frac{(t-1)^2(n-b)^2}{t^2}, \text{ and}$$
$$q_{22}(d) \leq n(t-1)\left(1 - \frac{1}{t} - \frac{1}{t^2}\right).$$

Hence, on simplification, for $b \in [0, n)$, from (2.3.9) it follows that

$$\begin{aligned} \operatorname{tr}(C_d) &\leq n(t-1) - \frac{2b}{t} - \frac{(t-1)^2(n-b)^2/t^2}{n(t-1)(1-1/t-1/t^2)} \\ &= n(t-1) - \frac{2b}{t} - \frac{(t-1)(n-b)^2}{n(t^2-t-1)}. \end{aligned} \tag{2.3.10}$$

Simple algebra shows that the right-hand side of (2.3.10) is maximized when $b = \frac{n}{t(t-1)}$. Recalling that $\frac{n}{t} \leq \frac{t-1}{2}$ and that b must be a nonnegative integer, it can be seen that the maximum value of the right-hand side of (2.3.10) is $n(t-1) - \frac{n(t-1)}{t^2-t-1}$. Hence the result. □

Hedayat and Yang (2004) observed that a balanced uniform design d^* with $p = t$ may not be universally optimal over $\Omega_{t,n,t}$ when $n > t(t-1)/2$. This will be evident when the reader reaches Chapter 3. For example, when $t = 3$, the design studied in Theorem 3.2.5 exists with $n = 36$ and is universally optimal for direct effects over $\Omega_{3,36,3}$, while a balanced uniform design which also exists in $\Omega_{3,36,3}$ (e.g., take six copies of the design d^* in Example 2.3.2) is *not* universally optimal. In order to study the performance of d^* for n beyond the range in Theorem 2.3.9, Hedayat and Yang (2004) extended this range for $t = 4, 5, \ldots, 12$. Their result is stated below.

Theorem 2.3.10. *Let $d^* \in \Omega_{t,n,t}$ be a balanced uniform design. Then d^* is universally optimal for direct effects over $\Omega_{t,n,t}$ for any $4 \leq t \leq 12$ when $n \leq t(t+2)/2$.* □

Remark 2.3.6. For $t = p = 4$, a balanced uniform design d^* exists with $n = 4, 8, 12$, etc., for example, design d_1 in Example 2.2.1 or copies of it. For $n = 4$, the optimality of d^* over $\Omega_{4,4,4}$ follows from Theorem 2.3.5 or Theorem 2.3.9. Theorem 2.3.10 establishes that the designs d^* with $n = 8$ and 12 are also universally optimal for direct effects in the entire class $\Omega_{4,8,4}$ and $\Omega_{4,12,4}$, respectively. For $t = p = 3$, Street, Eccleston and Wilson (1990) used a computer search to show that a balanced uniform design in $\Omega_{3,6,3}$ is A-optimal. The following result due to Hedayat and Yang (2004) shows that it is in fact universally optimal. □

Theorem 2.3.11. *A balanced uniform design $d^* \in \Omega_{3,6,3}$ is universally optimal.* □

For the case $n = 2t$, Theorems 2.3.9–2.3.11 together show that a balanced uniform design d^* is universally optimal for direct effects for all $t = p \geq 3$ over the general class $\Omega_{t,n,t}$. For larger n, these theorems, together with Theorem 2.3.8, establish that d^* is either universally optimal or highly efficient.

Now we consider the status of balanced uniform designs d^* for the estimation of carryover effects over the general class $\Omega_{t,n,p}$. Recall that Theorems 2.3.1–2.3.3 establish the universal optimality of d^* in this regard over restricted classes in $\Omega_{t,n=\mu_1 t,p}$. Kunert (1984b) showed that if μ_1 is even, d^* can never be universally optimal for carryover effects over $\Omega_{t,n=\mu_1 t,t}$ and if $\mu_1 = 1$, then d^* cannot be universally optimal over $\Omega_{t,n=t,t}$ if certain designs with some special properties exist. Instead, for $\mu_1 = t-1$, he gave the following result.

Theorem 2.3.12. *Let $n = t(t-1)$. An orthogonal residual effects design $f \in \Omega_{t,n,t}$ is universally optimal for the estimation of carryover effects over $\Omega_{t,n,t}$.*

Proof. The proof follows by noting that (i) $C_{f12} = \mathbf{0}$, (ii) C_{f22} is completely symmetric and, (iii) C_{f22} has the maximum trace over $\Omega_{t,n,t}$. □

2.4 Optimality of Strongly Balanced Designs

An interesting property of strongly balanced uniform designs is that for these designs, $C_{d12} = \mathbf{0}$ (see (2.2.2)), unlike balanced uniform designs. In view of this, expressions for C_d and $\bar{C}_d$ become particularly simple as shown in (2.2.3). This observation and the nice symmetry in the structure of these designs lead to strong optimality properties. These optimality results do not require any restriction on the class of competing designs, nor on n, as was required for balanced uniform designs in Section 2.3. The theorems presented in this section are due to Cheng and Wu (1980).

Theorem 2.4.1. *Let $d^* \in \Omega_{t,n,p}$ be a strongly balanced uniform design. Then d^* is universally optimal for the estimation of direct as well as carryover effects over $\Omega_{t,n,p}$.*

Proof. From (1.3.11), for any $d \in \Omega_{t,n,p}$, $C_d \leq C_{d11}$, and equality holds for $d = d^*$ since $C_{d^*12} = \mathbf{0}$ as shown in (2.2.2). Hence

$$\begin{aligned}\operatorname{tr}(C_d) &\leq \operatorname{tr}(C_{d11}) \leq \operatorname{tr}(C_{d^*11}), \text{ from Lemma 2.2.4} \\ &= \operatorname{tr}(C_{d^*}).\end{aligned}$$

Similarly,

$$\operatorname{tr}(\bar{C}_d) \leq \operatorname{tr}(C_{d22}) \leq \operatorname{tr}(C_{d^*22}) = \operatorname{tr}(\bar{C}_{d^*}).$$

Finally, from (2.2.3), C_{d^*} and $\bar{C}_{d^*}$ are both completely symmetric and the result follows by applying Theorem 1.4.1. □

Example 2.4.1. Let $t = 3, p = 6, n = 9$. The optimality of the design d_2 shown in Example 2.2.1 follows from Theorem 2.4.1. □

Remark 2.4.1. The optimal designs of Theorem 2.4.1 are rather large in size since for such a design to exist, we must necessarily have $t^2|n$, $t|p$ and $\mu_2 \geq 2$. By relaxing the condition of uniformity on subjects, one can obtain optimal designs that are smaller in size, as shown in the following result. □

Theorem 2.4.2. *Let $d^* \in \Omega_{t,n,p}$ be a strongly balanced design that is uniform on periods and uniform on subjects in the first $(p-1)$ periods. Then d^* is universally optimal for the estimation of direct as well as carryover effects over $\Omega_{t,n,p}$.*

Proof. We provide a proof for direct effects – the proof for carryover effects proceeds along similar lines. The existence of d^* in $\Omega_{t,n,p}$ implies that $n = \mu_1 t, p = \mu_4 t + 1$ for some integers $\mu_1, \mu_4 \geq 1$. Again, $Z_{d^*} = \mu_1\mu_4 J_t$, since d^* is strongly balanced. Moreover, since d^* is uniform on periods, from Lemma 2.2.1, $M_{d^*} = \mu_1 J_{tp}$, $\bar{M}_{d^*} = [\mathbf{0}_t \quad \mu_1 J_{t,p-1}]$, $\boldsymbol{r}_{d^*} = p\mu_1 \mathbf{1}_t$, $\bar{\boldsymbol{r}}_{d^*} = n\mu_4 \mathbf{1}_t$, and $\bar{N}_{d^*} = \mu_4 J_{tn}$ since d^* is uniform on subjects in the first $(p-1)$ periods. Hence, from (1.3.10) on simplification, $C_{d^*12} = \mathbf{0}$. Now, from (1.3.11), $C_{d^*} = C_{d^*11}$ and this is completely symmetric. The proof is complete as in Theorem 2.4.1 by invoking Lemma 2.2.5 and Theorem 1.4.1. □

Remark 2.4.2. The optimal design in Theorem 2.4.2 is the extra-period design of Patterson and Lucas (1959). □

Example 2.4.2. Let $t = 3, n = 6, p = 4$. The optimality of the design d_4 shown in Example 2.2.1 follows from Theorem 2.4.2. □

Cheng and Wu (1980) proved stronger optimality results for the optimal designs of Theorem 2.4.1 when the competing class of designs contains only *equireplicate* designs in $\Omega_{t,n,p}$. Let

$$\Omega_1 = \{d : d \in \Omega_{t,n,p}, \boldsymbol{r}_d = r\mathbf{1}_t \text{ for some integer } r \geq 1\}$$

and

$$\Omega_2 = \{d : d \in \Omega_{t,n,p}, \bar{\boldsymbol{r}}_d = \bar{r}\mathbf{1}_t \text{ for some integer } \bar{r} \geq 1\}.$$

Theorem 2.4.3. *Let $d^* \in \Omega_{t,n=\mu_1 t,p}$ be a strongly balanced uniform design. Then d^* minimizes the variance of the best linear unbiased estimator of any contrast among the direct effects (respectively, carryover effects) over Ω_1 (respectively, Ω_2).*

Proof. As before, we provide a proof for direct effects only, the proof for carryover effects is similar.

To prove the result for the direct effects, we need to prove that $C_{d^*} \geq C_d$ for all $d \in \Omega_1$. Recall that the row sums of C_d, $d \in \Omega_1$, are zero and thus, in particular, this is also true for C_{d^*}. So, it is enough to show that

$$\boldsymbol{x}'C_{d^*}\boldsymbol{x} \geq \boldsymbol{x}'C_d\boldsymbol{x} \text{ for all } d \in \Omega_1$$

and any $t \times 1$ vector $\boldsymbol{x}$ such that $\boldsymbol{x}'\mathbf{1}_t = 0$.

From (1.3.10) and (1.3.11), $C_d \leq C_{d11} \leq (np/t)I_t$ for any $d \in \Omega_1$, since by Lemma 2.2.2, $n^{-1}M_dM_d'$ and $p^{-1}N_dN_d' - (np)^{-1}N_dJ_nN_d'$ are both n.n.d. Thus we have

$$\boldsymbol{x}'C_d\boldsymbol{x} \leq \boldsymbol{x}'(\frac{np}{t}I_t)\boldsymbol{x} = \boldsymbol{x}'\frac{np}{t}(I_t - t^{-1}J_t)\boldsymbol{x} = \boldsymbol{x}'(\mu_1 pH_t)\boldsymbol{x} = \boldsymbol{x}'C_{d^*}\boldsymbol{x}$$

by (2.2.3). This completes the proof. □

2.5 Some More Optimal Designs

All the optimal designs for direct effects presented in Sections 2.3 and 2.4 are either balanced or strongly balanced. Kunert (1983) showed that certain designs which are neither balanced nor strongly balanced can also be universally optimal in the entire class of competing designs. In order to present his results, we need the following definitions.

Definition 2.5.1. A design $d \in \Omega_{t,n,p}$ will be called a balanced block design (BBD) on subjects if it satisfies the following conditions:
(i) d is equireplicate, i.e., r_{ds} is a constant for $1 \leq s \leq t$,
(ii) $T_d'\mathrm{pr}^{\perp}(U)T_d$ is completely symmetric, and

(iii) n_{dsj} is either $[p/t]$ or $[p/t]+1$, where $[x]$ is the greatest integer less than or equal to x, $1 \leq s \leq t$, $1 \leq j \leq n$.

Definition 2.5.2. A design $d \in \Omega_{t,n,p}$ will be called a BBD on periods if it satisfies the following conditions:
(i) d is equireplicate,
(ii) $T_d' \text{pr}^{\perp}(P) T_d$ is completely symmetric, and
(iii) m_{dsi} is either $[n/t]$ or $[n/t]+1$, $1 \leq s \leq t$, $1 \leq i \leq p$.

Definition 2.5.3. A design $d \in \Omega_{t,n,p}$ will be called a generalized Youden design (GYD) if d is a BBD on subjects and also a BBD on periods.

Definition 2.5.4. A design $d \in \Omega_{t,n,p}$ will be called a generalized Latin square (GLS) if d is a GYD, $t|n$ and $t|p$.

Clearly, any GYD as well as any GLS is equireplicate. We now present some results due to Kunert (1983).

Theorem 2.5.1. *Let t divide both n and p and suppose there is a GLS $d^* \in \Omega_{t,n,p}$ such that*

$$z_{d^*ss'} = \bar{r}_{d^*s'}/t, \; 1 \leq s, s' \leq t.$$

Then d^* is universally optimal for the estimation of direct effects over $\Omega_{t,n,p}$.

Proof. From the first form of C_d as given in (1.3.13), it can be shown as in (1.4.6) that

$$C_d \leq T_d' \text{pr}^{\perp}(\mathbf{1}_{np}) T_d$$

and as in (1.4.7), equality holds if and only if the following orthogonality condition is satisfied

$$T_d' \text{pr}^{\perp}(\mathbf{1}_{np})([P \;\; U \;\; F_d]) = \mathbf{0}. \tag{2.5.1}$$

The orthogonality condition (2.5.1) can be equivalently written as:

$$T_d' \text{pr}^{\perp}(\mathbf{1}_{np})([P \;\; U]) = \mathbf{0}, \tag{2.5.2}$$

and

$$T_d' \text{pr}^{\perp}(\mathbf{1}_{np}) F_d = \mathbf{0}. \tag{2.5.3}$$

Now $T_d' \text{pr}^{\perp}(\mathbf{1}_{np}) T_d$ is completely symmetric and has maximal trace if and only if d is equireplicate, a condition which is obviously met by the GLS d^*. Also, from the fact that d^* is a GLS, it is clear that it satisfies (2.5.2). Next, by (1.3.5), the condition (2.5.3) is equivalent to $Z_d = (np)^{-1} \boldsymbol{r}_d \bar{\boldsymbol{r}}_d'$, which, for equireplicate designs, reduces to $z_{dss'} = \bar{r}_{ds'}/t$, $1 \leq s, s' \leq t$. As specified in the statement of the theorem, d^* satisfies this condition. Hence the theorem follows. □

Remark 2.5.1. Note that for a GLS, $\bar{r}_{ds'}$ is a constant for $1 \leq s' \leq t$. Hence the condition on $z_{d^*ss'}$ to be satisfied by d^* in Theorem 2.5.1 simply means that d^* is also strongly balanced. In contrast, the following two results by Kunert (1983) demonstrate that non-uniform designs which are neither balanced nor strongly balanced can be universally optimal over $\Omega_{t,n,p}$. □

Theorem 2.5.2. *Suppose t divides p but not n and let there be a GYD $d^* \in \Omega_{t,n,p}$ such that*

$$z_{d^*ss'} = n^{-1}\left(\sum_{i=1}^{p} m_{d^*si}\bar{m}_{d^*s'i}\right), \quad 1 \leq s, s' \leq t.$$

Then d^* is universally optimal for the estimation of direct effects over $\Omega_{t,n,p}$.

Proof. As in the proof of Theorem 2.5.1, we have

$$C_d \leq T_d'\mathrm{pr}^{\perp}(P)T_d$$

and equality holds if and only if the following orthogonality condition is satisfied:

$$T_d'\mathrm{pr}^{\perp}(P)([U \quad F_d]) = \mathbf{0}. \tag{2.5.4}$$

The orthogonality condition (2.5.4) can be equivalently written as:

$$T_d'\mathrm{pr}^{\perp}(P)U = \mathbf{0}, \tag{2.5.5}$$

and

$$T_d'\mathrm{pr}^{\perp}(P)F_d = \mathbf{0}. \tag{2.5.6}$$

Now, $T_d'\mathrm{pr}^{\perp}(P)T_d$ is completely symmetric and has maximal trace if and only if d is BBD on periods. This is true for the design d^* in this theorem since it is a GYD. Also, from the fact that d^* is a GYD, it is clear that it satisfies (2.5.5). Finally, (2.5.6) is equivalent to the condition $Z_d = n^{-1}M_d\bar{M}_d'$, which, as specified in the statement of the theorem, is met by d^*. Hence the theorem follows. □

Theorem 2.5.3. *Suppose t divides n but not p and suppose there exists a GYD $d^* \in \Omega_{t,n,p}$ such that*

$$z_{d^*ss'} = p^{-1}\left(\sum_{j=1}^{n} n_{d^*sj}\bar{n}_{d^*s'j}\right), 1 \leq s, s' \leq t.$$

Then d^* is universally optimal for direct effects over $\Omega_{t,n,p}$. □

The proof of Theorem 2.5.3 is similar to that of Theorem 2.5.2 but is based on a different upper bound to C_d, namely, $T_d' \text{pr}^{\perp}(U) T_d$. Hence the orthogonality condition here is $T_d' \text{pr}^{\perp}(U)([P\ F_d]) = \mathbf{0}$ which is satisfied by d^* as in Theorem 2.5.3.

We display two designs in Example 2.5.1, both of which are universally optimal for direct effects over $\Omega_{t,n,p}$. These are neither balanced nor strongly balanced. The design d_2 satisfies the condition of Theorem 2.5.2 and is uniform on subjects but not on periods, whereas design d_1 satisfies the condition of Theorem 2.5.3 and is uniform on periods but not on subjects. The design d_1 is universally optimal for direct effects over $\Omega_{3,12,5}$ while d_2 is universally optimal over $\Omega_{3,4,15}$.

Example 2.5.1.

$$d_1 = \begin{matrix} 2\,1\,1 & 3\,1\,3 & 1\,2\,3 & 2\,2\,3 \\ 1\,2\,1 & 2\,3\,1 & 1\,3\,2 & 3\,2\,3 \\ 1\,2\,3 & 2\,3\,1 & 2\,3\,2 & 3\,1\,1 \\ 2\,1\,2 & 1\,1\,3 & 3\,1\,3 & 2\,3\,2 \\ 3\,3\,2 & 1\,2\,2 & 3\,1\,1 & 1\,3\,2 \end{matrix}, \quad d_2 = \begin{matrix} 1 & 1 & 2 & 3 \\ 2 & 3 & 1 & 2 \\ 3 & 2 & 3 & 1 \\ 2 & 2 & 3 & 1 \\ 3 & 1 & 1 & 2 \\ 1 & 1 & 2 & 3 \\ 2 & 3 & 2 & 1 \\ 3 & 3 & 1 & 2 \\ 1 & 2 & 3 & 3 \\ 1 & 2 & 1 & 3 \\ 2 & 2 & 1 & 3 \\ 3 & 1 & 3 & 2 \\ 1 & 1 & 3 & 2 \\ 2 & 3 & 2 & 1 \\ 3 & 3 & 2 & 1 \end{matrix}.$$

□

Remark 2.5.2. Kunert (1983) considered designs with a pre-period (i.e., designs where there are carryover effects in the first periods too) together with designs with no preperiods (i.e., no carryover effects in the first period) as we have considered so far. Some of his results, in particular the ones in Theorems 2.5.1–2.5.3, also hold over a class of competing designs that contains designs with pre-periods. Kunert (1983) also briefly considered the problem of finding designs that are optimal for carryover effects. We do not provide details on these and refer the reader to the original source for more information. Instead, we give an illustration in Example 2.5.2.

Example 2.5.2. Let $t = 3, n = 18, p = 6$. The following design is universally optimal for carryover effects over $\Omega_{3,18,6}$.

$$d \equiv \begin{matrix} 2\,3\,1 & 2\,3\,1 & 2\,3\,1 & 2\,3\,1 & 1\,2\,3 & 1\,2\,3 \\ 2\,3\,1 & 2\,3\,1 & 2\,3\,1 & 2\,3\,1 & 2\,3\,1 & 3\,1\,2 \\ 3\,1\,2 & 1\,2\,3 & 3\,1\,2 & 1\,2\,3 & 3\,1\,2 & 2\,3\,1 \\ 3\,1\,2 & 3\,1\,2 & 3\,1\,2 & 3\,1\,2 & 3\,1\,2 & 2\,3\,1 \\ 1\,2\,3 & 3\,1\,2 & 1\,2\,3 & 3\,1\,2 & 2\,3\,1 & 3\,1\,2 \\ 2\,3\,1 & 2\,3\,1 & 2\,3\,1 & 2\,3\,1 & 2\,3\,1 & 2\,3\,1 \end{matrix}.$$

□

Remark 2.5.3. Sometimes the values of t, n and p may be such that no GYD or GLS meeting the orthogonality conditions in the last three theorems exist. For such cases, Kunert (1983) discussed designs which nearly fulfill these conditions. As seen in Remark 2.4.1, $t^2|n$ is one of the necessary conditions for the existence of strongly balanced uniform designs. For n which is not a multiple of t^2, Kunert (1983) introduced the notion of nearly strongly balanced crossover designs as defined below. □

Definition 2.5.5. A design $d \in \Omega_{t,n,p}$ is said to be nearly strongly balanced if
(i) $Z_d Z_d'$ is completely symmetric, and
(ii) $z_{dss'}$ equals either $[n(p-1)/t^2]$ or $[n(p-1)/t^2]+1$ for all $1 \le s, s' \le t$.

We state the following results from Kunert (1983), whose proofs can be built essentially along the lines of those presented earlier in this section.

Theorem 2.5.4. *Let $n = at^2 + bt$, $1 \le b \le t-1$ and $p = \mu_2 t$ for some integers a, b and μ_2. Let $d^* \in \Omega_{t,n,p}$ be a nearly strongly balanced GLS. Then*
(i) d^* is universally optimal for the estimation of direct effects over the subclass of all designs in $\Omega_{t,n,p}$ which are uniform on subjects and the last period.

(ii) If in addition $a \geq b(t-b-1)t^{-1}$ *and* $\mu_2 \geq \max\{2, 4^{-1}b(t-b)+2t^{-1}\}$, *then* d^* *is universally optimal for the estimation of direct effects over the entire class* $\Omega_{t,n,p}$. □

Theorem 2.5.5. *Let* $n = at^2 + bt$, $1 \leq b \leq t-1$ *and* $p = \mu_2 t$ *for some integers* a, b *and* μ_2. *Let* $d^* \in \Omega_{t,n,p}$ *be a nearly strongly balanced GLS. Then* d^* *is universally optimal for the estimation of carryover effects over the subclass of designs in* $\Omega_{t,n,p}$ *which are uniform on the subjects and the first and last periods.* □

Example 2.5.3. The following two designs, $d_1 \in \Omega_{3,6,6}$ and $d_2 \in \Omega_{5,5,10}$, are nearly strongly balanced. By Theorem 2.5.4 (i) and Theorem 2.5.5, both designs are optimal over restricted classes in $\Omega_{t,n,p}$. However, unlike d_2, design d_1 also satisfies the condition of Theorem 2.5.4 (ii) and so it is universally optimal for direct effects over the entire class $\Omega_{3,6,6}$. Compared to a strongly balanced uniform design with the same t, both require fewer subjects.

$$d_1 = \begin{array}{cccccc} 1 & 2 & 3 & 1 & 2 & 3 \\ 2 & 3 & 1 & 1 & 2 & 3 \\ 3 & 1 & 2 & 2 & 3 & 1 \\ 3 & 1 & 2 & 3 & 1 & 2 \\ 2 & 3 & 1 & 3 & 1 & 2 \\ 1 & 2 & 3 & 2 & 3 & 1 \end{array}, \quad d_2 = \begin{array}{ccccc} 0 & 1 & 2 & 3 & 4 \\ 2 & 3 & 4 & 0 & 1 \\ 1 & 2 & 3 & 4 & 0 \\ 4 & 0 & 1 & 2 & 3 \\ 3 & 4 & 0 & 1 & 2 \\ 3 & 4 & 0 & 1 & 2 \\ 4 & 0 & 1 & 2 & 3 \\ 1 & 2 & 3 & 4 & 0 \\ 2 & 3 & 4 & 0 & 1 \\ 0 & 1 & 2 & 3 & 4 \end{array}.$$ □

Remark 2.5.4. If one desires universal optimality over the entire class $\Omega_{t,n,p}$, then with $t \leq 8$ and $b = t-1$, it follows that the smallest possible values of n and p for d^* as in Theorem 2.5.4 (ii) are $t^2 - t$ and $2t$, respectively (this happens for d_1 of Example 2.5.3 for $t = 3$). On the other hand, the smallest strongly balanced uniform design, which is also universally optimal over the entire class $\Omega_{t,n,p}$ (see Theorem 2.4.1), requires $n = t^2$ and $p = 2t$. Thus, for $t \leq 8$, the smallest design d^* as per Theorem 2.5.4 (ii) requires t fewer units than that required by a corresponding smallest strongly balanced uniform design while the number of periods is the same in both cases. However, when $t = 9$, the smallest value of p for d^* as per Theorem 2.5.4 (ii) is $3t$ and thus, the smallest d^* requires t more periods than the smallest strongly balanced uniform design. As t increases, this gap becomes wider. □

Bate and Jones (2006) showed that nearly strongly balanced uniform designs exist in $\Omega_{t,\mu_1 t,\mu_2 t}$ if and only if $\mu_1 = \alpha t + 1$ or, $\mu_1 = (\alpha + 1)t - 1$ for some integer $\alpha \geq 0$. So, the optimal designs of Theorems 2.5.4 and 2.5.5 can exist only for certain combinations of t, n and p. To obtain useful designs for more combinations of t, n, p, Bate and Jones (2006) defined a new class of designs, called nearly balanced uniform designs as follows.

Definition 2.5.6. A design $d \in \Omega_{t,n,p}$ is said to be a nearly balanced design if $z_{dss'}$ equals either $[n(p-1)/t^2]$ or $[n(p-1)/t^2] + 1$, for all $1 \leq s, s' \leq t$.

Thus, these nearly balanced designs satisfy condition (ii), but not necessarily condition (i), of Definition 2.5.5. As such, the information matrices C_d and $\bar{C}_d$ for these designs may not be completely symmetric and so, their universal optimality cannot be claimed via Theorem 1.4.1. However, if these designs are also uniform, then they are efficient for direct and carryover effects in the sense of Bate and Jones (2006), who called a design d^* to be efficient for direct (respectively, carryover) effects over a class if d^* maximizes $\text{tr}(C_d)$ (respectively, $\text{tr}(\bar{C}_d)$) over this class. Their result is given below.

Theorem 2.5.6. *Suppose* $n = at^2 + bt, 1 \leq b \leq t - 1$ *and* $p = \mu_2 t$ *for some integers* a, b, μ_2. *Let* d^* *be a nearly balanced uniform design in* $\Omega_{t,n,p}$. *Then*
(i) d^* *is efficient for the estimation of direct effects over the class of all designs in* $\Omega_{t,n,p}$ *which are uniform on subjects and the last period, and*
(ii) d^* *is efficient for the estimation of carryover effects over the class of all designs which are uniform on subjects and the first and the last periods.*

□

2.6 Constructions

In this section, we present some results on the construction of balanced and strongly balanced designs studied in Sections 2.3 and 2.4. We begin with the construction of balanced uniform designs. As stated in Section 2.2, the following conditions are necessary for the existence of these designs in $\Omega_{t,n,p}$:

$$\begin{aligned} &n = \mu_1 t \text{ for some positive integer } \mu_1, \\ &p = \mu_2 t \text{ for some positive integer } \mu_2, \text{ and} \\ &\mu_1(p-1) \equiv 0 \ (\text{mod } (t-1)). \end{aligned} \tag{2.6.1}$$

Moreover, if a balanced uniform design exists in $\Omega_{t,t,t}$, then by taking suitable copies of it we can obtain a balanced uniform design in $\Omega_{t,\mu_1 t,t}$. So, we first focus on the construction of such designs in $\Omega_{t,t,t}$.

For all *even* values of t, a balanced uniform design in $\Omega_{t,t,t}$ exists. This can be constructed (*cf.* Williams (1949)) as follows: Let

$$\boldsymbol{a}_0 = \left(1, t, 2, t-1, \ldots, \frac{t}{2} - 1, \frac{t}{2} + 2, \frac{t}{2}, \frac{t}{2} + 1\right)',$$

and for $1 \le u \le t-1$, define $\boldsymbol{a}_u = \boldsymbol{a}_0 + (u, \ldots, u)'$, where the entries in $\boldsymbol{a}_u$ are reduced modulo t and, thereafter, every 0 in $\boldsymbol{a}_u$ is replaced by t. Then the $t \times t$ array

$$A_t = [\boldsymbol{a}_0\ \boldsymbol{a}_1 \cdots\ \boldsymbol{a}_{t-1}] \tag{2.6.2}$$

is a balanced uniform crossover design in $\Omega_{t,t,t}$ with rows of A_t representing the periods and the columns, the subjects. Thus A_t as in (2.6.2) is simply a balanced Latin square and such a square is also referred to as a Williams square.

Example 2.6.1. Let $t = 4$. Here, $\boldsymbol{a}_0 = (1, 4, 2, 3)'$ and following the above construction method, the design d_1 of Example 2.2.1 may be constructed. This is a balanced uniform design in $\Omega_{4,4,4}$. □

Example 2.6.2. For $t = 6$, with $\boldsymbol{a}_0 = (1, 6, 2, 5, 3, 4)'$, we get the balanced uniform design in $\Omega_{6,6,6}$ as shown below:

$$\begin{array}{cccccc}
1 & 2 & 3 & 4 & 5 & 6 \\
6 & 1 & 2 & 3 & 4 & 5 \\
2 & 3 & 4 & 5 & 6 & 1 \\
5 & 6 & 1 & 2 & 3 & 4 \\
3 & 4 & 5 & 6 & 1 & 2 \\
4 & 5 & 6 & 1 & 2 & 3
\end{array}.$$
□

When t is *odd*, the following facts on the existence of a balanced uniform design in $\Omega_{t,t,t}$ are known.

(a) No such design exists for $t = 3, 5, 7$; see Laywine and Mullen (1998).

(b) Such designs for $t = 9, 15, 21$ and 27 were constructed by Archdeacon, Dinitz, Stinson and Tilson (1980), who called these squares row complete Latin squares (now also called Roman squares). The design for $t = 21$ was also constructed by Mendelsohn (1968) and is shown in full in Hedayat and Afsarinejad (1975). For $t = 39, 55, 57$, such squares can be constructed by using the methods of Mendelsohn (1968), Dénes and Keedwell (1974) and Wang (1973). The designs for $t = 9, 15$ and 27 are displayed in full by Hedayat and Afsarinejad (1978).

(c) Higham (1998) proved that a balanced uniform design exists in $\Omega_{t,t,t}$ when t is a composite number, i.e., it can be written as a product of two positive integers, each of which is greater than unity.

However, for *all odd* t, it has long been known (*cf.* Williams (1949), Sheehe and Bross (1961)) that a balanced uniform design exists in $\Omega_{t,2t,t}$. Such a design can be constructed as follows: Let

$$\boldsymbol{b}_0 = \left(1, t, 2, t-1, \ldots, \frac{t+5}{2}, \frac{t-1}{2}, \frac{t+3}{2}, \frac{t+1}{2}\right)',$$

$$\boldsymbol{c}_0 = \left(\frac{t+1}{2}, \frac{t+3}{2}, \frac{t-1}{2}, \frac{t+5}{2}, \ldots, t-1, 2, t, 1\right)',$$

i.e., the column $\boldsymbol{c}_0$ is obtained by writing the entries of $\boldsymbol{b}_0$ in the reverse order. For $1 \le u \le t-1$, let $\boldsymbol{b}_u = \boldsymbol{b}_0 + (u, \ldots, u)'$ and $\boldsymbol{c}_u = \boldsymbol{c}_0 + (u, \ldots, u)'$, where the elements of $\boldsymbol{b}_u$ and $\boldsymbol{c}_u$ are reduced modulo t and, thereafter, every 0 therein is replaced by t. Then a balanced uniform design in $\Omega_{t,2t,t}$ is given by the $t \times 2t$ array

$$B_t = [\boldsymbol{b}_0\ \boldsymbol{b}_1 \cdots \boldsymbol{b}_{t-1}\ \boldsymbol{c}_0\ \boldsymbol{c}_1 \cdots \boldsymbol{c}_{t-1}]. \tag{2.6.3}$$

Example 2.6.3. Let $t = 5$. Then

$$\boldsymbol{b}_0 = (1, 5, 2, 4, 3)' \text{ and } \boldsymbol{c}_0 = (3, 4, 2, 5, 1)'.$$

Following the method of construction just described, a balanced uniform design in $\Omega_{5,10,5}$ can be obtained, which is shown below.

1	2	3	4	5	3	4	5	1	2
5	1	2	3	4	4	5	1	2	3
2	3	4	5	1	2	3	4	5	1 .
4	5	1	2	3	5	1	2	3	4
3	4	5	1	2	1	2	3	4	5

□

Stufken (1996) proved the following result regarding the existence of a balanced uniform design in $\Omega_{t,\mu_1 t,\mu_2 t}$.

Theorem 2.6.1. *The necessary conditions (2.6.1) for the existence of balanced uniform designs in $\Omega_{t,n,p}$ are also sufficient if t is even. Furthermore, for odd t, a balanced uniform design exists in $\Omega_{t,n,p}$ if, in addition to the necessary conditions, n/t is even.*

Proof. Let the necessary conditions in (2.6.1) hold and let t be even. Define $W = (w_{ij})$ to be a $\mu_2 \times \mu_1$ matrix with entries from the set $\{1, 2, \ldots, t\}$, such that among the differences $w_{ij} - w_{i-1,j}$, $2 \le i \le \mu_2$, $1 \le j \le \mu_1$, reduced modulo t, each of the numbers $0, 1, \ldots, \frac{1}{2}t - 1, \frac{1}{2}t + 1, \ldots, t - 1$ occurs equally often. As observed by Stufken (1996), such a matrix W exists for all positive integers μ_1, μ_2 and t even. With A_t as defined in (2.6.2), for $1 \le u \le t - 1$, let A_u be obtained by adding u to each element of A_t, addition being performed modulo t, and then every 0 in A_u being replaced by t. Consider the $\mu_2 t \times \mu_1 t$ array U given by

$$U = \begin{bmatrix} A_{w_{11}} & \cdots & A_{w_{1\mu_1}} \\ \vdots & & \\ A_{w_{\mu_2 1}} & \cdots & A_{w_{\mu_2 \mu_1}} \end{bmatrix}. \tag{2.6.4}$$

It can then be verified that U is a balanced uniform design in $\Omega_{t,\mu_1 t,\mu_2 t}$, where t is even.

If t is odd and μ_1 is even, construct a $\mu_2 \times \mu_1/2$ matrix $F = (f_{ij})$ with entries from $\{1, 2, \ldots, t\}$, such that among the differences $f_{ij} - f_{i-1,j}$, $2 \le i \le \mu_2$, $1 \le j \le \mu_1/2$, the $(t-1)/2$ numbers $(t+3)/2, (t+7)/2, \ldots, (t-3)/2$, reduced modulo t, occur equally often. Again, as noted by Stufken (1996), such a matrix F exists. With the $t \times 2t$ matrix B_t defined in (2.6.3), for $1 \le u \le t - 1$, let B_u be obtained by adding u to each element of B_t, addition being modulo t, and then every 0 in B_u being replaced by t. The $\mu_2 t \times \mu_1 t$ matrix V given by

$$V = \begin{bmatrix} B_{f_{11}} & \cdots & B_{f_{1(\mu_1/2)}} \\ \vdots & & \\ B_{f_{\mu_2 1}} & \cdots & B_{f_{\mu_2(\mu_1/2)}} \end{bmatrix} \tag{2.6.5}$$

is a balanced uniform design in $\Omega_{t,\mu_1 t,\mu_2 t}$, where t is odd and μ_1 is even. □

However, the condition that μ_1 is even when t is odd is not a necessary condition for the existence of these designs in $\Omega_{t,n=\mu_1 t,t}$. This is evident from the construction given by Prescott (1999), who proved that a balanced uniform design exists for odd $t > 3$ in $\Omega_{t,3t,t}$. His construction involves three initial treatment sequences given by

$$\left(0, 1, (t-2), 3, \ldots, \frac{t+(-1)^{(t-1)/2}}{2}, \frac{t+(-1)^{(t-1)/2}}{2}+1, \ldots, 4, (t-1), 2\right)',$$

$$\left(0, 2, (t-1), 4, \ldots, \frac{t+(-1)^{(t-1)/2}}{2}+1, \frac{t+(-1)^{(t-1)/2}}{2}, \ldots, 3, (t-2), 1\right)',$$

$$\left(0, (t-1), (t-2), 2, (t-4), 4, \ldots, \frac{t-2-(-1)^{(t-1)/2}}{2}, \ldots, 3, (t-5), 1, (t-3)\right)'.$$

In the third sequence, we begin with 0 and $(t-1)$. Thereafter, we work from each end in pairs, entering $(t-2), (t-3), \ldots$ and $1, 2, \ldots,$ alternately, until all the labels are used. The required design can be constructed by developing these three initial sequences mod t. In the following example, we show a balanced uniform design in $\Omega_{5,15,5}$, following Prescott's construction.

Example 2.6.4. Let $t = 5$. Then the following is a balanced uniform design in $\Omega_{5,15,5}$ (here the treatment symbols are $0, 1, \ldots, 4$):

0	1	2	3	4	0	1	2	3	4	0	1	2	3	4
1	2	3	4	0	2	3	4	0	1	4	0	1	2	3
3	4	0	1	2	4	0	1	2	3	3	4	0	1	2 .
4	0	1	2	3	3	4	0	1	2	1	2	3	4	0
2	3	4	0	1	1	2	3	4	0	2	3	4	0	1

□

We now take up the construction of strongly balanced uniform designs. The earliest examples of strongly balanced uniform designs were given by Quenouille (1953), who displayed these designs for $t = 3, n = 18, p = 6$ and $t = 4, n = 16, p = 8$. Berenblut (1964) and Patterson (1973) gave general methods of construction of strongly balanced uniform designs in $\Omega_{t,t^2,2t}$. Cheng and Wu (1980) constructed such designs where $t^2|n$ and p/t is an even integer; d_2 in Example 2.2.1 is one such design. It may be noted that if a strongly balanced uniform design in $\Omega_{t,t^2,2t}$, as constructed by the method of Cheng and Wu (1980), is represented by a $2t \times t^2$ array, A, then one can obtain a strongly balanced uniform design in $\Omega_{t,\alpha t^2,2\beta t}$, where α, β are integers. This design is given by the $2\beta t \times \alpha t^2$ array B given below, where as usual, the rows of B represent the periods and columns, the subjects. In B, the array A appears α times in each row and β times in each column as follows:

$$B = \begin{matrix} A\,A \cdots A \\ A\,A \cdots A \\ \vdots \\ A\,A \cdots A \end{matrix}.$$

Recall from Section 2.2 that a set of necessary conditions for the existence of a strongly balanced uniform design are that (i) $t^2|n$, (ii) $t|p$, and (iii) $p \geq 2t$. Therefore, for the existence of such designs, we need that

$$n = \mu_3 t^2, \text{ for some integer } \mu_3 \geq 1,$$
$$p = \mu_2 t, \text{ for some integer } \mu_2 \geq 2.$$

Sen and Mukerjee (1987) gave a construction of these designs for odd μ_2 (see Section 6.3) and this, together with the construction of Cheng and Wu (1980) for μ_2 even, shows that the above necessary conditions are sufficient as well. Stufken (1996) gave a unified method of construction of these designs for general μ_2, which covers both the odd and even cases, and we give his construction in Theorem 2.6.2. His construction uses an *orthogonal array* of strength *two*, which is defined below. For a comprehensive account of orthogonal arrays, we refer to Hedayat, Sloane and Stufken (1999).

Definition 2.6.1. An orthogonal array, $OA(n,p,t,2)$ of strength two is a $p \times n$ array with entries from a set of t symbols, such that any $2 \times n$ subarray contains each ordered pair of symbols equally often, precisely n/t^2 times.

Remark 2.6.1. From Definition 2.6.1 it is clear that n must be a multiple of t^2. Moreover, a necessary condition for the existence of an $OA(t^2,p,t,2)$ is that $p \leq t+1$. In particular, an $OA(t^2, t+1, t, 2)$ exists whenever t is a prime or a prime power (see, e.g., Chapters 2 and 3 in Hedayat *et al.* (1999)). □

Example 2.6.5. An $OA(9,3,3,2)$ is shown below.

$$\begin{bmatrix} 1 & 1 & 1 & 2 & 2 & 2 & 3 & 3 & 3 \\ 1 & 2 & 3 & 1 & 2 & 3 & 1 & 2 & 3 \\ 1 & 2 & 3 & 2 & 3 & 1 & 3 & 1 & 2 \end{bmatrix}.$$

□

Theorem 2.6.2. *A strongly balanced uniform design exists whenever* $n = \mu_3 t^2$ *and* $p = \mu_2 t$, *for integers* $\mu_3 \geq 1$ *and* $\mu_2 \geq 2$.

Proof. Consider an orthogonal array $A_0 \equiv OA(t^2, 3, t, 2)$, where the symbols of the array are $1, 2, \ldots, t$. It is well known that such an array exists for all $t \geq 2$. Let B_0 be an orthogonal array $OA(t^2, 2, t, 2)$, obtained from A_0 by deleting its third row. For $1 \leq u \leq t-1$, let A_u be a $3 \times t^2$ matrix obtained by adding u to each element of A_0 and similarly, let B_u be a $2 \times t^2$ matrix obtained by adding u to each element of B_0, where the elements of A_u and B_u are reduced modulo t, and then every 0 therein is replaced by t. Finally, let A and B be the $3t \times t^2$ and $2t \times t^2$ matrices defined as

$$A = \begin{bmatrix} A_0 \\ A_1 \\ \vdots \\ A_{t-1} \end{bmatrix}, B = \begin{bmatrix} B_0 \\ B_1 \\ \vdots \\ B_{t-1} \end{bmatrix}.$$

Since $\mu_2 \geq 2$, let $\mu_2 = 3\alpha + 2\beta$ for some nonnegative integers α, β. It can then be verified that the $p \times t^2$ array

$$[A' \cdots A' \; B' \cdots B']'$$

consisting of α copies of A and β copies of B is a strongly balanced uniform design in $\Omega_{t,t^2,p=\mu_2 t}$. Now juxtaposing copies of this design, we get a strongly balanced uniform design in $\Omega_{t,n=\mu_3 t^2,p=\mu_2 t}$. □

Example 2.6.6. Let $t = 3, n = 9, p = 6$. Using the orthogonal array shown in Example 2.6.5 as A_0 we have

$$B_0 = \begin{bmatrix} 1\,1\,1\,2\,2\,2\,3\,3\,3 \\ 1\,2\,3\,1\,2\,3\,1\,2\,3 \end{bmatrix}.$$

Since $\mu_2 = 2$, we take $\alpha = 0, \beta = 1$. Now, forming B_1 and B_2 as in the proof of Theorem 2.6.2, we get the strongly balanced uniform design $d_2 \in \Omega_{3,9,6}$ of Example 2.2.1. □

Example 2.6.7. Let $t = 3$ and we now take $n = 9 = p$. Here, $\mu_2 = 3$ and thus, $\alpha = 1, \beta = 0$. With A_0 as in Example 2.6.5 and forming A_1 and A_2 as indicated in the proof of Theorem 2.6.2, we can construct the following strongly balanced uniform design in $\Omega_{3,9,9}$.

$$\begin{matrix} 1 & 1 & 1 & 2 & 2 & 2 & 3 & 3 & 3 \\ 1 & 2 & 3 & 1 & 2 & 3 & 1 & 2 & 3 \\ 1 & 2 & 3 & 2 & 3 & 1 & 3 & 1 & 2 \\ 2 & 2 & 2 & 3 & 3 & 3 & 1 & 1 & 1 \\ 2 & 3 & 1 & 2 & 3 & 1 & 2 & 3 & 1 \\ 2 & 3 & 1 & 3 & 1 & 2 & 1 & 2 & 3 \\ 3 & 3 & 3 & 1 & 1 & 1 & 2 & 2 & 2 \\ 3 & 1 & 2 & 3 & 1 & 2 & 3 & 1 & 2 \\ 3 & 1 & 2 & 1 & 2 & 3 & 2 & 3 & 1 \end{matrix}.$$ □

Theorem 2.6.2 shows that the necessary conditions for the existence of a strongly balanced uniform design are also sufficient. An alternative proof of this was given by Sen and Mukerjee (1987) (see Remark 6.3.4).

Finally, we consider the construction of strongly balanced designs which are uniform on periods and uniform on subjects in the first $p-1$ periods. Cheng and Wu (1980) observed that such designs can be easily constructed in $\Omega_{t,n,t+1}$ if a balanced uniform design in $\Omega_{t,n,t}$ is available. This can be achieved by simply repeating the tth period of the balanced uniform design as the $(t+1)$th period to get the desired strongly balanced design. Such designs are called extra-period designs.

Example 2.6.8. Let $t = 4 = n, p = 5$. From the design d_1 in Example 2.2.1, we get the following strongly balanced design in $\Omega_{4,4,5}$ which is uniform on subjects and uniform on the first 4 periods.

$$\begin{array}{cccc} 1 & 2 & 3 & 4 \\ 4 & 1 & 2 & 3 \\ 2 & 3 & 4 & 1 \\ 3 & 4 & 1 & 2 \\ 3 & 4 & 1 & 2 \end{array}.$$ □

Stufken (1996) showed that this idea of Cheng and Wu (1980) can be extended to obtain such strongly balanced designs even for $p > t+1$. His result is given below.

Theorem 2.6.3. *If $n = \mu_1 t$ and $p = \mu_4 t + 1$, μ_1, μ_4 being positive integers, then there exists a strongly balanced crossover design in $\Omega_{t,n,p}$ which is uniform on periods and uniform on subjects in the first $(p-1)$ periods if (i) t is even, or (ii) t is odd and μ_1 is even.*

Proof. For t even, let the $t \times t$ arrays $A_1, \ldots, A_t$ be as in the proof of Theorem 2.6.1. Construct a $\mu_4 t \times \mu_1 t$ array as follows: first juxtapose μ_1 copies of A_t to get the first t rows, then juxtapose μ_1 copies of $A_{t/2}$ to get the next t rows and continue like this, alternating between A_t and $A_{t/2}$, until $\mu_4 t$ rows are obtained. Repeat the last row of the resultant array as the $(\mu_4 t + 1)$th row. It can be verified that this $(\mu_4 t + 1) \times \mu_1 t$ array is a strongly balanced crossover design in $\Omega_{t,\mu_1 t,\mu_4 t+1}$ which is uniform on periods and uniform on subjects in the first $\mu_4 t$ periods.

If t is odd and μ_1 is even, let B_t be as in (2.6.3). Construct a $\mu_4 t \times \mu_1 t$ array as follows: first juxtapose $\mu_1/2$ copies of B_t to get the first t rows. Next, again juxtapose $\mu_1/2$ copies of B_t to get the next t rows but this time, permute the columns within each copy of B_t such that the first row of this juxtaposition is identical to the last row of the previous juxtaposition. Continue this process until $\mu_4 t$ rows are obtained, ensuring that while juxtaposing, the copies of B_t are such that the periods it and $it + 1$, $1 \leq i \leq \mu_4 - 1$ are identical. Finally, repeat the last row of the resultant array as the $(\mu_4 t + 1)$th row. This $(\mu_4 t + 1) \times \mu_1 t$ array gives the required design. □

Example 2.6.9. We illustrate the method of construction of Theorem 2.6.3 for t even. Let $t = 4, n = 8, p = 9$. Here, $\mu_1 = 2, \mu_4 = 2$ and the array A_4 is given by the design d_1 of Example 2.2.1. Proceeding as in the proof of Theorem 2.6.3, we next consider the 8×8 array

$$\begin{bmatrix} A_4 \ A_4 \\ A_2 \ A_2 \end{bmatrix},$$

and finally, repeating the last row, we get a strongly balanced design in $\Omega_{4,8,9}$, which is shown below.

$$\begin{array}{cccccccc} 1 & 2 & 3 & 4 & 1 & 2 & 3 & 4 \\ 4 & 1 & 2 & 3 & 4 & 1 & 2 & 3 \\ 2 & 3 & 4 & 1 & 2 & 3 & 4 & 1 \\ 3 & 4 & 1 & 2 & 3 & 4 & 1 & 2 \\ 3 & 4 & 1 & 2 & 3 & 4 & 1 & 2 \\ 2 & 3 & 4 & 1 & 2 & 3 & 4 & 1 \\ 4 & 1 & 2 & 3 & 4 & 1 & 2 & 3 \\ 1 & 2 & 3 & 4 & 1 & 2 & 3 & 4 \\ 1 & 2 & 3 & 4 & 1 & 2 & 3 & 4 \end{array}.$$ □

Example 2.6.10. We now illustrate the method of construction of Theorem 2.6.3 for t odd. Let $t = 3, n = 6, p = 7$. Here, $\mu_1 = 2, \mu_4 = 2$. Following the construction method described in the proof of Theorem 2.6.3, we get a strongly balanced design in $\Omega_{3,6,7}$, which is shown below.

$$\begin{array}{cccccc} 1 & 2 & 3 & 2 & 3 & 1 \\ 3 & 1 & 2 & 3 & 1 & 2 \\ 2 & 3 & 1 & 1 & 2 & 3 \\ 2 & 3 & 1 & 1 & 2 & 3 \\ 1 & 2 & 3 & 2 & 3 & 1 \\ 3 & 1 & 2 & 3 & 1 & 2 \\ 3 & 1 & 2 & 2 & 3 & 1 \end{array}.$$ □

Methods of construction of nearly balanced and nearly strongly balanced designs have been given by Bate and Jones (2006) and we refer the reader to the original source for details on these. In Example 2.5.3, we have exhibited two nearly strongly balanced uniform designs. The next example shows a nearly balanced uniform design as given by Bate and Jones (2006).

Example 2.6.11. Let $t = 5, n = 10, p = 15$. The following is a nearly balanced uniform design in $\Omega_{5,10,15}$. Here, the treatment labels are 0, 1, 2, 3 and 4.

$$\begin{array}{cccccccccc}
0 & 1 & 2 & 3 & 4 & 2 & 3 & 4 & 0 & 1 \\
2 & 3 & 4 & 0 & 1 & 1 & 2 & 3 & 4 & 0 \\
1 & 2 & 3 & 4 & 0 & 4 & 0 & 1 & 2 & 3 \\
4 & 0 & 1 & 2 & 3 & 3 & 4 & 0 & 1 & 2 \\
3 & 4 & 0 & 1 & 2 & 3 & 4 & 0 & 1 & 2 \\
3 & 4 & 0 & 1 & 2 & 4 & 0 & 1 & 2 & 3 \\
4 & 0 & 1 & 2 & 3 & 1 & 2 & 3 & 4 & 0 \\
1 & 2 & 3 & 4 & 0 & 2 & 3 & 4 & 0 & 1 \\
2 & 3 & 4 & 0 & 1 & 0 & 1 & 2 & 3 & 4 \\
0 & 1 & 2 & 3 & 4 & 0 & 1 & 2 & 3 & 4 \\
0 & 1 & 2 & 3 & 4 & 0 & 1 & 2 & 3 & 4 \\
1 & 2 & 3 & 4 & 0 & 4 & 0 & 1 & 2 & 3 \\
4 & 0 & 1 & 2 & 3 & 1 & 2 & 3 & 4 & 0 \\
2 & 3 & 4 & 0 & 1 & 3 & 4 & 0 & 1 & 2 \\
3 & 4 & 0 & 1 & 2 & 2 & 3 & 4 & 0 & 1
\end{array} .$$

□

Chapter 3

Some Optimal Designs with $p < t$

3.1 Introduction

In Chapter 2, results on the optimality of crossover designs have been presented for the situation when the number of periods is at least equal to the number of treatments, i.e., when $p \geq t$. However, in practice, it is sometimes difficult to continue experiments over a large number of periods and so, one looks for designs where the number of periods is small compared to the number of treatments, i.e., $p < t$. Patterson (1952) was probably the first to provide several methods of construction for balanced crossover designs with $p \leq t$. Examples of crossover experiments with $p < t$ were also given by Freeman (1959). Several others including Patterson and Lucas (1962), Atkinson (1966), Hedayat and Afsarinejad (1975), Constantine and Hedayat (1982), Afsarinejad (1983) and Stufken (1991) constructed designs with $p \leq t$. In this chapter, we consider some optimal designs for this situation. Further results, also for $p < t$, but under settings different from that in this chapter, will be presented in some later chapters.

The problem of obtaining optimal crossover designs with $p < t$ was first considered by Dey, Gupta and Singh (1983), who proved the universal optimality of certain balanced designs. However, the class of competing designs considered by them was rather restricted. Subsequently, several results on optimal crossover designs for the situation $p \leq t$ have been obtained. We present some of this work and begin with results due to Afsarinejad (1985) and Stufken (1991) in Section 3.2. We also describe in Section 3.2 some optimality results given by Hedayat, Stufken and Yang (2006) under a model in which the subject effects are treated as random. Optimality results on two-period crossover designs due to Hedayat and Zhao (1990) and Carriere and Reinsel (1993) are discussed in Section 3.3. In Section 3.4, some results obtained by Shah, Bose and Raghavarao (2005) on the optimality aspects of a class of balanced designs studied by Patterson (1952) are reviewed. Finally, in Section 3.5, some construction methods of the designs of Patterson (1952) are discussed. We use notation as in Chapter 1. The following definition will be helpful.

Definition 3.1.1. A $u \times v$ array having entries from a set of $t \geq 2$ distinct symbols is called a type I orthogonal array of strength two if in any $2 \times v$ subarray all $t(t-1)$ ordered two-tuples without repetition occur equally often.

A type I orthogonal array of strength two will be denoted by $OA_I(v, u, t, 2)$. Clearly, a necessary condition for the existence of an $OA_I(v, u, t, 2)$ is that $v = \lambda t(t-1)$ for some positive integer λ. Two such type I orthogonal arrays are shown below.

Example 3.1.1.

$$OA_I(6,3,3,2) = \begin{bmatrix} 0\,1\,2\,0\,1\,2 \\ 1\,2\,0\,2\,0\,1 \\ 2\,0\,1\,1\,2\,0 \end{bmatrix},$$

$$OA_I(12,4,4,2) = \begin{bmatrix} 1\,1\,1\,2\,2\,2\,3\,3\,3\,4\,4\,4 \\ 2\,3\,4\,1\,3\,4\,1\,2\,4\,1\,2\,3 \\ 4\,2\,3\,3\,4\,1\,4\,1\,2\,2\,3\,1 \\ 3\,4\,2\,4\,1\,3\,2\,4\,1\,3\,1\,2 \end{bmatrix}.$$

□

Remark 3.1.1. Rao (1961) showed that an orthogonal array $OA(t^2, k, t, 2)$ (see Definition 2.6.1) can be used to construct a type I orthogonal array $OA_I(t(t-1), k-1, t, 2)$. Therefore, from Remark 2.6.1, it follows that an $OA_I(t(t-1), p, t, 2)$ exists for any $p \leq t$ if t is a prime or a prime power. □

3.2 Designs with $p \leq t$

All the results discussed in this section are under the traditional model (1.2.1). Afsarinejad (1985) considered a class $\mathcal{D}$ of designs in $\Omega_{t,n,p}$ with $t \geq 3$, $n = 2t$ and $p = (t+1)/2$, such that for every $d \in \mathcal{D}$, (i) d is uniform on periods, (ii) the subjects of d form the blocks of a balanced incomplete block (BIB) design with block size p, and (iii) when restricted to the first $p-1$ periods, the subjects of d again form the blocks of a BIB design with block size $p - 1$. He showed that a balanced design in $\mathcal{D}$ is universally optimal for the estimation of direct and carryover effects over $\mathcal{D}$. Since the class $\mathcal{D}$ is somewhat restrictive, we do not discuss the details of this result here. Afsarinejad (1985) also provided a construction method of these universally optimal designs in $\mathcal{D}$ when t is a prime or a prime power of the form $4\alpha + 3$, for some nonnegative integer α. We give an example below with $t = 7$, where the treatments are labeled as $0, 1, \ldots, 6$.

Example 3.2.1.

$$\begin{array}{cccccccccccccc} 1 & 4 & 3 & 0 & 5 & 6 & 2 & 6 & 1 & 5 & 3 & 4 & 0 & 2 \\ 2 & 5 & 4 & 1 & 6 & 0 & 3 & 5 & 0 & 4 & 2 & 3 & 6 & 1 \\ 4 & 0 & 6 & 3 & 1 & 2 & 5 & 3 & 5 & 2 & 0 & 1 & 4 & 6 \\ 0 & 3 & 2 & 6 & 4 & 5 & 1 & 0 & 2 & 6 & 4 & 5 & 1 & 3 \end{array} . \qquad \square$$

It is easy to verify that the design in Example 3.2.1 belongs to $\mathcal{D}$ and is also balanced. By the result of Afsarinejad (1985), this design is universally optimal for direct and carryover effects over $\mathcal{D}$. Note that when $t = 4\alpha+3$ is a prime number, such designs were constructed earlier by Patterson (1952); see Theorem 3.5.3.

Stufken (1991) relaxed the restrictions on the competing designs as imposed by Afsarinejad (1985) and proposed designs which he showed to be universally optimal for either direct effects or carryover effects in different subclasses of $\Omega_{t,n,p}$. His results are given in Theorems 3.2.1–3.2.4. For proving his results, Stufken (1991) gave an interesting set of conditions under which the information matrices of direct and carryover effects (i.e., C_d and $\bar{C}_d$) reduce to simple forms which are completely symmetric. These in turn help in obtaining optimal (or, highly efficient) designs in a large class of competing designs. In order to present these results, we need the following notation.

Using the notation as introduced in Chapter 1, for a design $d \in \Omega_{t,n,p}$, let

$$n^i_{dsj} = \begin{cases} 1, \text{ if } d(i,j) = s, \\ 0, \text{ otherwise,} \end{cases}$$

and, for $1 \leq s, s' \leq t; s \neq s'$, define

$$\begin{aligned} S_1 &= z_{dss'} - n^{-1} \sum_{i=2}^{p} m_{dsi} m_{ds',i-1}, \\ S_2 &= \sum_{j=1}^{n} (n^p_{dsj} - n^{-1} m_{dsp})(\bar{n}_{ds'j} - n^{-1} \bar{r}_{ds'}), \\ S_3 &= \sum_{j=1}^{n} (\bar{n}_{dsj} - n^{-1} \bar{r}_{ds})(\bar{n}_{ds'j} - n^{-1} \bar{r}_{ds'}) + n^{-1} \sum_{i=1}^{p-1} m_{dsi} m_{ds'i}. \end{aligned} \qquad (3.2.1)$$

The following result due to Stufken (1991) is then true.

Theorem 3.2.1. *Let $d \in \Omega_{t,n,p}$ be a design which is uniform on periods and for which the quantities S_1, S_2 and S_3 defined in (3.2.1) are independent of s, s'; $1 \le s, s' \le t$, $s \neq s'$. Then the information matrices C_d and $\bar{C}_d$ are completely symmetric and are given by*

$$C_d = (tp)^{-1}\left[np(p-1) + t^2(2S_2 + S_3) - \{t^2(S_2+S_3) - t^2pS_1 - n(p-1)\}^2/\delta_1\right] H_t,$$

$$\bar{C}_d = (tp)^{-1}\left[n(p-1)^2 + t^2S_3 - \{t^2(S_2+S_3) - t^2pS_1 - n(p-1)\}^2/\delta_2\right] H_t, \tag{3.2.2}$$

where $\delta_1 = n(p-1)^2 + t^2S_3$, $\delta_2 = np(p-1) + t^2(2S_2+S_3)$ and as usual, $H_t = I_t - t^{-1}J_t$. □

Remark 3.2.1. Several designs discussed in Chapter 2 satisfy the conditions of Theorem 3.2.1. We give some illustrations below, which, incidentally, show that these conditions can be met by designs with $p \le t$ as well as by designs with $p > t$.

(i) If $d \in \Omega_{t,n,p}$ is a balanced uniform design then from Definitions 2.2.1 and 2.2.4, $m_{dsi} = n/t$ and $z_{dss'} = \lambda_1 = n(p-1)/\{t(t-1)\}$ for all $s \neq s', i$, $1 \le s, s' \le t$; $1 \le i \le p$. Consequently, on simplification from (3.2.1),

$$S_1 = n(p-1)/\{t^2(t-1)\}.$$

Again, every treatment occurs exactly $\mu_1(= n/t)$ times in the last period, and since d is also uniform on subjects, if any treatment s appears in the last period on subject j then, for this subject, $n^p_{dsj} = 1$ while $\bar{n}_{dsj} = (p/t) - 1$ and $\bar{n}_{ds'j} = (p/t)$ for all $s' \neq s$. Hence, on simplification from (3.2.1) we have

$$S_2 = \mu_1\left(1 - \frac{1}{n}\frac{n}{t}\right)\left(\frac{p}{t} - \frac{p-1}{t}\right) + \mu_1\left(-\frac{1}{t}\right)\left(\frac{p}{t} - 1 - \frac{p-1}{t}\right) + \left(n - \frac{2n}{t}\right)\left(-\frac{1}{t}\right)\left(\frac{p}{t} - \frac{p-1}{t}\right) = \frac{n}{t^2},$$

$$S_3 = 2\mu_1\left(\frac{p}{t} - 1 - \frac{p-1}{t}\right)\left(\frac{p}{t} - \frac{p-1}{t}\right) + \left(n - \frac{2n}{t}\right)\left(\frac{p}{t} - \frac{p-1}{t}\right)^2 + \mu_1\left(\frac{p-1}{t}\right) = \frac{n(p-2)}{t^2},$$

and thus S_1, S_2, S_3 are all independent of $s, s', s \neq s'$. Consequently, by Theorem 3.2.1, C_d and $\bar{C}_d$ are completely symmetric and on simplification from (3.2.2) using S_1, S_2 and S_3 as obtained above, C_d and $\bar{C}_d$ reduce to the expressions as given by C_{d_3} and $\bar{C}_{d_3}$, respectively, in (2.2.3).

(ii) If $d \in \Omega_{t,n,p}$ is a strongly balanced uniform design then, from Definitions 2.2.1 and 2.2.5, $m_{dsi} = n/t$ as in (i) above and $z_{dss'} = n(p-1)/t^2$ for all s, s'. For this case, (3.2.1) yields $S_1 = 0$ and since d is uniform over subjects, S_2 and S_3 are as in (i). Hence, from Theorem 3.2.1, C_d and $\bar{C}_d$ are completely symmetric and their expressions in (3.2.2) reduce to those for C_{d_4} and $\bar{C}_{d_4}$, respectively, in (2.2.3).

(iii) The optimal designs with $t = 4\alpha + 3, n = 2t$ and $p = (t+1)/2$ as constructed by Afsarinejad (1985) also satisfy the conditions of Theorem 3.2.1.

(iv) Consider a crossover design $d \in \Omega_{t,n,p}$, given by an orthogonal array $OA(n, p, t, 2)$ (*cf.* Definition 2.6.1) with the t symbols, n columns and p rows of the orthogonal array representing, respectively, the t treatments, n subjects and p periods of d. It can be checked that this design also satisfies the conditions of Theorem 3.2.1 with $S_u = 0$ for all $u = 1, 2, 3$. For example, the fact that

$$\sum_{j=1}^{n} \bar{n}_{dsj}\bar{n}_{ds'j} = \frac{n}{t^2}(p-1)(p-2) \quad \text{for } s \neq s',$$

arising from the properties of an orthogonal array of strength two, is crucial in showing that $S_3 = 0$. Hence, these designs have completely symmetric information matrices C_d and $\bar{C}_d$. For the case $p = 2$, Hedayat and Zhao (1990) showed that such a design is universally optimal for direct effects over $\Omega_{t,n,2}$ (see Remark 3.3.3 below).

(v) Consider a crossover design $d \in \Omega_{t,n=t(t-1),p}$ given by an orthogonal array of type I, $OA_I(t(t-1), p, t, 2)$ (*cf.* Definition 3.1.1) with its symbols as the t treatments, rows as the p periods and columns as the n subjects. From Definition 3.1.1 it can be checked that this design is uniform on periods, balanced, and satisfies the conditions of Theorem 3.2.1 with $S_1 = S_2 = t^{-1}(p-1)$ and $S_3 = t^{-1}(p-1)(p-2)$. Thus, these designs too have completely symmetric information matrices C_d and $\bar{C}_d$. □

Remark 3.2.2. Stufken (1991) pointed out that if there are two crossover designs $d_1 \in \Omega_{t,n_1,p}$ and $d_2 \in \Omega_{t,n_2,p}$, both satisfying the conditions of Theorem 3.2.1, then the juxtaposed design $d = [d_1\ d_2] \in \Omega_{t,n_1+n_2,p}$ also satisfies these conditions. Furthermore, for this design d, each S_u is equal to the sum of the S_u values corresponding to d_1 and d_2, $1 \leq u \leq 3$. This fact is quite useful in the search for optimal crossover designs and will be used in Theorem 3.2.2 below. □

Suppose t and p are such that an $OA_I(t(t-1), p, t, 2)$ exists. Stufken (1991) suggested the following construction method of a crossover design with t treatments, p periods and $n = \lambda t(t-1)$ subjects, where λ is a positive integer. Consider the following $p \times t(t-1)$ arrays:

A_1 : Any $OA_I(t(t-1), p, t, 2)$;

A_2 : Any $OA_I(t(t-1), p-1, t, 2)$ forms the first $p-1$ rows and the $(p-1)$th row is repeated as row p;

A_3 : Any $OA_I(t(t-1), p-1, t, 2)$ forms the first $p-1$ rows and any row other than the $(p-1)$th is repeated as row p, $p \geq 3$.

Let $0 \leq c_0 \leq c_1 \leq \lambda$ and let $\mathcal{D}^* \subset \Omega_{t,\lambda t(t-1),p}$ be the class of all crossover designs which are constructed by juxtaposing $\lambda - c_1$ copies of A_1, c_0 copies of A_2 and $c_1 - c_0$ copies of A_3. Using the observation made in Remark 3.2.2, it can be seen that a design $d \in \mathcal{D}^*$ satisfies the conditions of Theorem 3.2.1 with

$$S_1 = \lambda(p-1)/t - c_0,\ S_2 = \lambda(p-1)/t - c_1,\ S_3 = \lambda(p-1)(p-2)/t. \quad (3.2.3)$$

It now remains to find c_0 and c_1 appropriately to arrive at an optimal design. The following results were obtained by Stufken (1991) in this context.

Theorem 3.2.2. *Let $d^* \in \mathcal{D}^*$ be a design for which $c_0 = c_1 = c^*$, where c^* is the nearest integer to $\{t(p-1)\}^{-1}\lambda$ or either of the two nearest integers if there is a tie. Then d^* is universally optimal for direct effects over $\mathcal{D}^*$.*

Proof. By Theorem 3.2.1, for any design $d \in \mathcal{D}^*$, the information matrix for direct effects C_d is as given in (3.2.2) with the S_u's as in (3.2.3), $1 \leq u \leq 3$. Since C_d is completely symmetric, in order to prove the result by invoking Theorem 1.4.1, it is enough to maximize the multiplier of H_t in C_d over c_0, c_1, where $0 \leq c_0 \leq c_1 \leq \lambda$. After substituting the expressions for S_u's in C_d, it can be seen that this maximization is achieved when $c_0 = c_1$ and their common value is as given in the theorem. □

Theorem 3.2.3. *For $p \leq t$, let $d^* \in \mathcal{D}^*$ be a design for which $c_0 = c_1 = \lambda$. Then d^* is universally optimal for carryover effects over the entire class $\Omega_{t,\lambda t(t-1),p}$.*

Proof. Clearly, d^* only consists of λ copies of A_2. From the structure of A_2 it can be shown that $C_{d^*12} = \mathbf{0}$. Moreover, d^* maximizes $\mathrm{tr}(C_{d22})$ over $\Omega_{t,\lambda t(t-1),p}$ and hence from (2.2.3), d^* also maximizes $\mathrm{tr}(\bar{C}_d)$ over

$\Omega_{t,\lambda t(t-1),p}$. Again, from Theorem 3.2.1, $\bar{C}_d$ is completely symmetric and the proof is completed by invoking Theorem 1.4.1. □

Example 3.2.2. Let $t = 4, p = 3, n = 12$. The optimal design of Theorem 3.2.3 consists of one copy of A_2 which has as its first two rows an $OA_I(12, 2, 4, 2)$. Therefore, repeating the second row, the optimal design is

$$\begin{array}{cccccccccccc} 1 & 1 & 1 & 2 & 2 & 2 & 3 & 3 & 3 & 4 & 4 & 4 \\ 2 & 3 & 4 & 1 & 3 & 4 & 1 & 2 & 4 & 1 & 2 & 3 \\ 2 & 3 & 4 & 1 & 3 & 4 & 1 & 2 & 4 & 1 & 2 & 3 \end{array}.$$ □

Remark 3.2.3. For $p = t$, the result in Theorem 3.2.3 reduces to a result of Kunert (1984b) (see Theorem 2.3.12). □

Stufken (1991) also proposed another class of designs and showed that these are optimal for direct effects in a subclass of $\Omega_{t,n,p}$ which is wider than the class $\mathcal{D}^*$ of Theorem 3.2.1. We call these designs "Stufken designs" and define them as follows.

Definition 3.2.1. A design $d \in \Omega_{t,n,p}$ is said to be a Stufken design if it satisfies the following conditions:

(a) d is uniform on periods.

(b) d, when restricted to the first $p-1$ periods and viewed as a block design with the subjects as blocks, is a BIB design with n blocks, each of size $p-1$.

(c) In the last period of d, θ subjects receive a treatment that was not assigned to them in any of the previous periods, while the other subjects receive the same treatment as in period $p-1$, where θ is the nearest integer (or one of the nearest integers) to $n(pt-t-1)/\{(p-1)t\}$.

(d) $z_{dss'} - p^{-1} \sum_{j=1}^{n} n_{dsj}\bar{n}_{ds'j}$, $s \neq s'$, is independent of s and s'.

(e) $\sum_{j=1}^{n} n_{dsj}n_{ds'j}$, $s \neq s'$, is independent of s and s'.

Theorem 3.2.4 below gives the optimality properties of these designs as proved by Stufken (1991). Later, Kushner (1998) showed that if n is divisible by $t(p-1)$, then the Stufken designs are universally optimal in $\Omega_{t,n,p}$; we will come back to this point in Remark 4.5.4. Subsequently, Hedayat and Yang (2004) further extended this result and proved that if a Stufken design exists in $\Omega_{t,n,p}$, it is indeed universally optimal for direct effects over the entire class $\Omega_{t,n,p}$, even without the aforesaid divisibility condition; see Theorem 3.2.5. To give a flavor of the ideas involved, we indicate the proof of Theorem 3.2.4 as given by Stufken (1991). The proof of the result of Hedayat and Yang (2004) is more involved; we therefore omit it and refer to the original paper for the proof.

Let $\mathcal{D}^0_{t,n,p}$ be the class of all crossover designs in $\Omega_{t,n,p}$ for which condition (b) of Definition 3.2.1 holds.

Theorem 3.2.4. *Let $d^* \in \Omega_{t,n,p}$ be a Stufken design. Then d^* is universally optimal for direct effects over $\mathcal{D}^0_{t,n,p}$.*

Proof. For any design, let

θ_1 : the number of subjects that receive a treatment in period p that was not assigned to them in any of the previous periods,

θ_2 : the number of subjects that receive the same treatment in period p as in period $p-1$.

Stufken (1991) showed that for any $d \in \mathcal{D}^0_{t,n,p}$,

$$\operatorname{tr}(C_d) \leq n(p-1) - \frac{2(n-\theta_1)}{p} - \frac{pt}{n(p-1)(pt-t-1)}(n-\theta_2-p^{-1}\theta_1)^2, \tag{3.2.4}$$

with equality if d satisfies conditions (a) and (d) of Definition 3.2.1.

Clearly, $n \geq \theta_1 + \theta_2$ and in the above expression, $(n - \theta_2 - p^{-1}\theta_1) \geq 0$. So, for fixed θ_1, the right-hand side of (3.2.4) will be maximized when θ_2 is as large as possible, i.e., when $\theta_2 = n - \theta_1$. Putting this value of θ_2 in the right-hand side of (3.2.4), one gets the following expression

$$n(p-1) - 2\frac{(n-\theta_1)}{p} - \theta_1^2\frac{t(p-1)}{np(pt-t-1)}$$

as a function of θ_1. It is easily verified that this expression is maximized when θ_1 is equal to the nearest integer to $n(pt-t-1)/\{(p-1)t\}$, i.e., when (c) holds. Let this optimum value of θ_1 be denoted by θ_1^*.

Thus, any design satisfying the conditions (a)–(d) maximizes $\operatorname{tr}(C_d)$ over $\mathcal{D}^0_{t,n,p}$ and so,

$$\operatorname{tr}(C_{d^*}) = n(p-1) - \frac{2(n-\theta_1^*)}{p} - \frac{(p-1)t}{np(pt-t-1)}\theta_1^{*2} = \max_{d \in \mathcal{D}^0_{t,n,p}} \operatorname{tr}(C_d). \tag{3.2.5}$$

Since d^* also satisfies condition (e), C_{d^*} is completely symmetric, and the result now follows on invoking Theorem 1.4.1. □

When a Stufken design exists in $\Omega_{t,n,p}$, Hedayat and Yang (2004) established that the maximization in (3.2.5) is valid over all $d \in \Omega_{t,n,p}$ and not only over $\mathcal{D}^0_{t,n,p}$. Their result is stated in the following theorem and its proof is omitted here.

Theorem 3.2.5. *Let $d^* \in \Omega_{t,n,p}$ be a Stufken design. Then d^* is universally optimal for direct effects over $\Omega_{t,n,p}$.* □

Remark 3.2.4. Consider the optimal design d^* described in Theorem 3.2.2 when $\lambda = at(t-1)$ or equivalently, $n = at^2(t-1)^2$ and $p = t$ for some integer a. Then $c^* = \{t(p-1)\}^{-1}\lambda = a$ and this optimal design is the one with $a\{t(t-1)-1\}$ copies of A_1 and a copies of A_2. By Theorem 3.2.2, this design is optimal over $\mathcal{D}^*$. Moreover, for this design $\theta_1 = n - at(t-1) = \theta_1^*$ and so, by Theorem 3.2.5, it is also universally optimal in $\Omega_{t,n,t}$. □

Remark 3.2.5. Let $n \leq t(p-1)/2$. Then $\theta_1^* = n$. Thus for $n = 2t$ and $p = (t+1)/2$, the designs considered by Afsarinejad (1985) are universally optimal for direct effects over $\mathcal{D}^0_{t,n,p}$ if $t \geq 11$. If $p = t$ and $n \leq \frac{1}{2}t(t-1)$, then also $\theta_1^* = n$ and in this case, Theorem 3.2.4 reduces to a result given by Kunert (1984b), which was later extended by Hedayat and Yang (2003); see Remark 2.3.5. However, note that in this case, universal optimality holds in the entire class $\Omega_{t,n,t}$ as shown in Theorem 2.3.9. □

We now describe some results under a model in which the subject effects are random. These results are due to Hedayat *et al.* (2006). A model with random subject effects, in contrast to the fixed effects model (1.2.1), is justified in situations where the subjects included in the experiment can be viewed as representing a random sample from a population of subjects. The model considered by Hedayat *et al.* (2006) is the same as (1.2.1), *except* that now the subject effects are considered as uncorrelated random variables with mean zero and variance σ^2_β. Also, the random subject effects and the errors are assumed to be mutually uncorrelated. Thus, following the notation introduced in Chapter 1, we may write the model as

$$\begin{aligned}\mathbb{E}(\boldsymbol{Y}_d) &= \mu\mathbf{1}_{np} + P\boldsymbol{\alpha} + T_d\boldsymbol{\tau} + F_d\boldsymbol{\rho},\\ \mathbb{D}(\boldsymbol{Y}_d) &= \sigma^2\Sigma = \sigma^2(I_n \otimes V),\end{aligned} \tag{3.2.6}$$

where $V = (I_p + \gamma J_p)$, $\gamma = \sigma^2_\beta/\sigma^2$ and σ^2, as before, is the error variance.

Then C_d, the information matrix for direct effects under model (3.2.6) may be written analogous to (1.3.18) as

$$C_d = T_d'(I_n \otimes V^{-\frac{1}{2}})\text{pr}^\perp\{(I_n \otimes V^{-\frac{1}{2}})([\mathbf{1}_{np}\; P\; F_d])\}(I_n \otimes V^{-\frac{1}{2}})T_d.$$

As usual, for any d, an upper bound for this C_d is given by the information matrix for direct effects under a model ignoring period effects. Using this fact and a generalization of Lemma 1.4.2, Hedayat *et al.* (2006) obtained an upper bound for $\mathrm{tr}(C_d)$ in a form similar to (2.3.9) and showed that this bound is attained by a design $d \in \Omega_{t,n,p}$ if the following conditions hold:

$$\begin{aligned}&\text{(a) } T_d'P = p^{-1}\mathbf{r}_d\mathbf{1}_p'.\\ &\text{(b) Each of the matrices: } T_d'(I_n \otimes V^{-\frac{1}{2}})\mathrm{pr}^{\perp}(\mathbf{1}_{np})(I_n \otimes V^{-\frac{1}{2}})T_d,\\ &\qquad T_d'(I_n \otimes V^{-\frac{1}{2}})\mathrm{pr}^{\perp}(\mathbf{1}_{np})(I_n \otimes V^{-\frac{1}{2}})F_d, \text{ and}\\ &\qquad F_d'(I_n \otimes V^{-\frac{1}{2}})\mathrm{pr}^{\perp}(\mathbf{1}_{np})(I_n \otimes V^{-\frac{1}{2}})F_d\\ &\text{is completely symmetric.}\end{aligned} \tag{3.2.7}$$

Let $\Omega_{t,n,p}^* \subset \Omega_{t,n,p}$ be the class consisting of designs in which each treatment is replicated n/t times in the last period and in which no treatment is preceded by itself in any subject. Within this class, Hedayat *et al.* (2006) identified a design d^* for which C_{d^*} is completely symmetric, its trace is equal to the upper bound referred to above, and which maximizes this upper bound over all $d \in \Omega_{t,n,p}^*$. We state their result below.

Theorem 3.2.6. *Suppose $p \le t$ and let $d^* \in \Omega_{t,n,p}$ be a design which*
(i) satisfies conditions (a) and (b) in (3.2.7),
(ii) is equireplicate, and
(iii) is binary over subjects, i.e., each treatment appears at most once in each subject.
Then, under model (3.2.6), irrespective of the value of $\gamma(= \sigma_\beta^2/\sigma^2)$, d^* is universally optimal under model (3.2.6) for direct effects over $\Omega_{t,n,p}^*$. □

Later in Chapter 5, we will introduce totally balanced designs in Definition 5.2.3. It will be evident from this definition that any such design with $p \le t$ satisfies all the conditions of Theorem 3.2.6 and is thus universally optimal for direct effects over $\Omega_{t,n,p}^*$, no matter what the value of γ is. As an illustration of Theorem 3.2.6, Hedayat *et al.* (2006) gave the following design d, which is universally optimal for direct effects over $\Omega_{5,20,3}^*$ for any value of γ.

Example 3.2.3.

$$d = \begin{matrix} 1\,2\,3\,4\,5\,4\,3\,2\,1\,5\,1\,3\,5\,2\,4\,2\,5\,3\,1\,4 \\ 5\,1\,2\,3\,4\,5\,4\,3\,2\,1\,4\,1\,3\,5\,2\,4\,2\,5\,3\,1 \\ 2\,3\,4\,5\,1\,3\,2\,1\,5\,4\,3\,5\,2\,4\,1\,5\,3\,1\,4\,2 \end{matrix}.$$

□

Theorem 3.2.6 is particularly interesting since it identifies universally optimal designs irrespective of the value of γ. However, the optimal designs of this theorem are in general not optimal over the entire class $\Omega_{t,n,p}$ and their efficiency over $\Omega_{t,n,p}$ depends on the value of γ. For additional results and details, the original source may be consulted.

3.3 Two-period Designs

In this section we consider an important particular case of the general situation $p < t$, namely, the case where $p = 2$, which is the minimum number of periods required to ensure the estimability of the direct effect contrasts. Two-period designs are of substantial interest in the field of clinical trials and for a discussion on the design and analysis of two-period crossover experiments and their application in clinical trials see, e.g., Grizzle (1965), Hills and Armitage (1979), Armitage and Hills (1982) and Willan and Pater (1986). See also Balaam (1968) and Barker, Hews, Huitson and Poloniecki (1982). In this section, we review some results on the optimality of two-period crossover designs as obtained by Hedayat and Zhao (1990) and Carriere and Reinsel (1993). Further optimal designs with two periods, which were studied by Afsarinejad and Hedayat (2002) under a model with self and mixed carryover effects, are presented in Chapter 5.

Throughout this section, $p = 2$. Let d be a design in $\Omega_{t,n,2}$ and for $1 \leq s \leq t$, let f_{ds} and g_{ds} denote the numbers of times treatment s appears in the first period and in the second period, respectively, of d. Thus in the notation of Chapter 1, $z_{ds's}$ is the number of subjects that receive treatment s in the first period and treatment s' in the second period and clearly, as in (1.3.2),

$$f_{ds} = \bar{r}_{ds} = \sum_{s'=1}^{t} z_{ds's}, \quad \sum_{s'=1}^{t} z_{dss'} = g_{ds}, \quad \sum_{s=1}^{t} f_{ds} = \sum_{s=1}^{t} g_{ds} = n. \qquad (3.3.1)$$

In their study of optimal designs with $p = 2$, Hedayat and Zhao (1990) assumed the model (1.3.14) with an error covariance structure $\Sigma = I_n \otimes V$, where

$$V = \begin{pmatrix} 1 & \rho \\ \rho & 1 \end{pmatrix}.$$

As noted in Lemma 1.3.1, for $p = 2$, the information matrices C_d and $\bar{C}_d$ under model (1.3.14) are equal to a constant times the corresponding information matrices under the traditional model (1.3.3), this constant being a function of ρ. In view of this fact, in searching for optimal designs in $\Omega_{t,n,2}$ under model (1.3.14), we may as well work with the expressions of the information matrices under (1.3.3). So, henceforth, we proceed with the model (1.3.3). Then C_d and $\bar{C}_d$, are as given by (1.3.11) with $p = 2$.

Hedayat and Zhao (1990) proved their optimality results by making a very interesting connection between a two-period crossover design in $\Omega_{t,n,2}$ and a block design. This connection is shown in Theorem 3.3.1; it makes the study of optimality of two-period crossover designs simpler as the optimality of block designs is well studied. Theorems 3.3.1–3.3.5 are all due to Hedayat and Zhao (1990). Let $\mathcal{C}_d$ denote the information matrix for treatments for an arbitrary block design d under the usual additive linear model for block designs.

Theorem 3.3.1. *Let d be a design in $\Omega_{t,n,2}$ and let there be b distinct treatments in the first period of d. Let these treatments be labeled as $1, 2, \ldots, b$, where $b \leq t$. Then there exists a block design d_0 with t treatments and b blocks of sizes $\bar{r}_{d1}, \ldots, \bar{r}_{db}$, such that the (s', s)th element of the treatment versus block incidence matrix of d_0 equals $z_{ds's}$ $(1 \leq s' \leq t, 1 \leq s \leq b)$, and the relationship*

$$\mathcal{C}_{d_0} = 2C_d, \tag{3.3.2}$$

holds. Conversely, from a block design with t treatments and $b(\leq t)$ blocks one can obtain a crossover design d in $\Omega_{t,n,2}$, with n equal to the total number of experimental units in the block design, such that (3.3.2) holds.

Proof. Given a design $d \in \Omega_{t,n,2}$ with b distinct treatments in the first period, we construct a block design d_0 with b blocks as follows: Corresponding to treatment s, $1 \leq s \leq b$, we look at all subjects j for which treatment s appears in period 1 of d. Then the treatments which appear in the second period of d for these subjects will together constitute block s of the design d_0. Hence, each of the distinct treatments in period 1 of d gives rise to one block of d_0 and thus d_0 is obtained.

Conversely, given a block design d_0 with b blocks and $t(\geq b)$ treatments, we may construct a design $d \in \Omega_{t,n,2}$ as follows: Suppose the size of the sth block of d_0 is k_s. We first assign the treatment s to k_s subjects in the first period of d. Then the k_s treatments appearing in block s of d_0 are assigned to these k_s subjects in the second period, $1 \leq s \leq b$. This gives the design d in $\Omega_{t,n,2}$.

It can be checked that (3.3.2) holds by computing $\mathcal{C}_{d_0}$ and comparing it with the C_d as obtained from (1.3.11) with $p = 2$. Alternatively, one can see the truth of (3.3.2) by noting that under model (1.3.3), since subject effects are fixed, the best linear unbiased estimators of estimable contrasts among direct effects depend only on the quantities $Y_{2j} - Y_{1j}$, $1 \le j \le n$, as one has to eliminate the subject effects. Thus, for $p = 2$, one can remodel the $2n$ observations in $\boldsymbol{Y}_d$ in model (1.3.3) via the n terms $\{Y_{2j} - Y_{1j}\}$ as follows:

$$\mathbb{E}(Y_{2j} - Y_{1j}) = \alpha_2 - \alpha_1 + \tau_{d(2,j)} + \rho_{d(1,j)} - \tau_{d(1,j)},\ 1 \le j \le n, \qquad (3.3.3)$$

with $\text{Var}(Y_{2j} - Y_{1j}) = 2\sigma^2$ and $\text{Cov}(Y_{2j} - Y_{1j}, Y_{2j'} - Y_{1j'}) = 0$, $j \ne j', 1 \le j, j' \le n$. Observe that (3.3.3) is equivalent to the usual additive linear model for a block design, where $\alpha_2 - \alpha_1$ is identified with the general mean, the quantities $\rho_{d(1,j)} - \tau_{d(1,j)}$ are identified with the block effects and the quantities $\tau_{d(2,j)}$ are identified with the treatment effects, such that each "observation" arising from the block design has variance $2\sigma^2$. Hence (3.3.2) is evident. □

Remark 3.3.1. Had we worked with the correlated errors model (1.3.14) instead of (1.3.3), the only change in the above proof would have been that we would have obtained $\text{Var}(Y_{2j} - Y_{1j}) = 2(1 - \rho)\sigma^2$. □

Remark 3.3.2. In view of (3.3.2), a crossover design d is connected for direct effects if and only if the corresponding block design d_0 is connected. But in d_0, treatment s is replicated g_{ds} times, and hence d_0 is connected only if $g_{ds} > 0$ for every s. Hence, in order to ensure connectedness for direct effects, hereafter we consider only those two-period crossover designs in which every treatment is applied at least once in the second period. □

The next example illustrates the connection between d and d_0 as shown in Theorem 3.3.1.

Example 3.3.1. Let $t = 3, n = 9$ and consider a design $d \in \Omega_{3,9,2}$ given by

$$d = \begin{matrix} 1 & 1 & 1 & 2 & 2 & 3 & 3 & 3 & 3 \\ 1 & 2 & 3 & 1 & 1 & 1 & 1 & 2 & 3 \end{matrix}.$$

Following the construction method given in the proof of Theorem 3.3.1, the corresponding block design d_0 is given by the blocks:

$$\begin{matrix} \text{Block I}: & 1\ 2\ 3 \\ \text{Block II}: & 1\ 1 \\ \text{Block III}: & 1\ 1\ 2\ 3 \end{matrix}.$$

It can be verified that $C_{d_0} = 2C_d$.

Now, consider the following block design d_0 involving 5 treatments and 4 blocks:

$$\begin{array}{r} \text{Block I}: \ 1\ 2\ 3\ 4 \\ \text{Block II}: \ 1\ 2\ 3 \\ \text{Block III}: 1\ 2\ 3\ 4\ 5 \\ \text{Block IV}: 2\ 3\ 4\ 5 \end{array}.$$

The corresponding crossover design $d \in \Omega_{5,16,2}$ is shown below:

$$d = \begin{array}{cccccccccccccccc} 1 & 1 & 1 & 1 & 2 & 2 & 2 & 3 & 3 & 3 & 3 & 3 & 4 & 4 & 4 & 4 \\ 1 & 2 & 3 & 4 & 1 & 2 & 3 & 1 & 2 & 3 & 4 & 5 & 2 & 3 & 4 & 5 \end{array}.$$

Once again, one can verify that $C_{d_0} = 2C_d$. The labeling of the treatments and blocks is arbitrary. □

We also note from Theorem 3.3.1 that if d_0 is a proper block design with b blocks of equal size, then in the corresponding crossover design d, the b treatments appearing in the first period do so equally often. In view of this observation, we have the following corollary.

Corollary 3.3.1. *If a block design is optimal in the class of all proper block designs with t treatments and b blocks, where $b \leq t$, then the corresponding crossover design is also optimal for direct effects in the class of all crossover designs in which b treatments appear in the first period and do so equally often.* □

Example 3.3.2. $t = 3, n = 12, p = 2$. Consider the following two designs studied by Kunert (1983).

$$d_1 = \begin{array}{cccccccccccc} 1 & 2 & 3 & 2 & 3 & 1 & 1 & 2 & 3 & 2 & 3 & 1 \\ 2 & 3 & 1 & 1 & 2 & 3 & 2 & 3 & 1 & 1 & 2 & 3 \end{array},$$

$$d_2 = \begin{array}{cccccccccccc} 1 & 2 & 3 & 1 & 2 & 3 & 1 & 2 & 3 & 2 & 3 & 1 \\ 1 & 2 & 3 & 2 & 3 & 1 & 1 & 2 & 3 & 1 & 2 & 3 \end{array}.$$

Kunert (1983) showed by actual computation that $C_{d_2} > C_{d_1}$. Hedayat and Zhao (1990) showed that d_2 is universally optimal in the class of those designs in $\Omega_{3,12,2}$ which are uniform on the first period. This follows from Corollary 3.3.1 because the block design corresponding to d_2 is a balanced block design (BBD) and hence, is universally optimal in the class of proper block designs. □

In view of Corollary 3.3.1, one can obtain optimal two-period crossover designs by searching for an optimal block design with corresponding parameters. The following result is a step towards that direction.

Theorem 3.3.2. *Let d_1 be a block design with t treatments and $b \leq t$ blocks and let d_2 be the design with a single block consisting of all the blocks of d_1 taken together. Then $C_{d_1} \leq C_{d_2}$, with equality if and only if, for every fixed s, the frequency of treatment s in block u of d_1 is proportional to the size of the uth block, $1 \leq s \leq t, 1 \leq u \leq b$.*

Proof. It is well known that the condition of proportional frequencies is equivalent to the condition of orthogonality between treatments and blocks, see, e.g., Lemma 2.1 of Dey (1986). Now, the result is immediate. □

It follows then that a block design in which each treatment appears in each block equally often is universally optimal in the class of all proper block designs with t treatments and an arbitrary number of blocks.

Theorem 3.3.3. *A design $d^* \in \Omega_{t,n,2}$ with $n \equiv 0 \pmod t$ is universally optimal for direct effects over $\Omega_{t,n,2}$ if and only if*
(a) $f_{d^*s} = 0 \pmod t$, $1 \leq s \leq t$, and
(b) $z_{d^*s's} = t^{-1} f_{d^*s}$, $1 \leq s' \leq t$.

Proof. If a block design d_0^* is obtained from d^* as in Theorem 3.3.1, then condition (b) guarantees that each treatment occurs an equal number of times in each block of d_0^*. Now the result follows by invoking Theorems 3.3.1 and 3.3.2. □

Remark 3.3.3. The number of distinct treatments in the first period of the design d^* in Theorem 3.3.3 may be any number b with $b \leq t$ and by condition (a), f_{d^*s} is a multiple (possibly zero) of t, $1 \leq s \leq t$, subject to $\sum_{s=1}^{t} f_{d^*s} = n$. Moreover, from condition (b), it is not hard to see that the quantities g_{d^*s}, $1 \leq s \leq t$, are all equal and so d^* is uniform on the second period. In particular, if a design is given by an $OA(n, 2, t, 2)$ then by Theorem 3.3.3, its universal optimality follows. □

Example 3.3.3. Suppose $t = 3$. Hedayat and Zhao (1990) showed that by Theorem 3.3.3, each of the following four designs is universally optimal for direct effects over $\Omega_{3,12,2}$.

$$d_1 = \begin{matrix} 1\,1\,1\,1\,1\,1\,1\,1\,1\,1\,1\,1 \\ 1\,2\,3\,1\,2\,3\,1\,2\,3\,1\,2\,3 \end{matrix}, \quad d_2 = \begin{matrix} 1\,1\,1\,1\,1\,1\,1\,1\,1\,2\,2\,2 \\ 1\,2\,3\,1\,2\,3\,1\,2\,3\,1\,2\,3 \end{matrix},$$

$$d_3 = \begin{matrix} 1\,1\,1\,1\,1\,1\,2\,2\,2\,2\,2\,2 \\ 1\,2\,3\,1\,2\,3\,1\,2\,3\,1\,2\,3 \end{matrix}, \quad d_4 = \begin{matrix} 1\,1\,1\,1\,1\,1\,2\,2\,2\,3\,3\,3 \\ 1\,2\,3\,1\,2\,3\,1\,2\,3\,1\,2\,3 \end{matrix}.$$

Note that $C_{d_1} = C_{d_2} = C_{d_3} = C_{d_4}$ and $b = 1$ for d_1, $b = 2$ for d_2 and d_3

and $b = 3$ for d_4. So, no other design which is non-isomorphic to either of d_u, $1 \leq u \leq 4$, is optimal in $\Omega_{3,12,2}$. □

Example 3.3.4. Suppose $t = 3$. Then the following design represented by the $OA(9, 2, 3, 2)$ is universally optimal for direct effects in $\Omega_{3,9,2}$.

$$\begin{matrix} 1 & 1 & 1 & 2 & 2 & 2 & 3 & 3 & 3 \\ 1 & 2 & 3 & 1 & 2 & 3 & 1 & 2 & 3 \end{matrix} .$$ □

By Theorem 3.3.3, one may construct several optimal designs in $\Omega_{t,n,2}$. To decide which one of these to use in practice, Hedayat and Zhao (1990) recommended that we use the design which gives the largest degrees of freedom for estimating σ^2. This is achieved if the design assigns a single treatment to all the subjects in the first period; e.g., the design d_1 in Example 3.3.3. However, this design will not allow the estimation of contrasts among carryover effects. If these estimates are desired, then one has to use designs where all treatments are allocated in the first period; e.g., the design d_4 in Example 3.3.3.

Next, we consider the case when n is not a multiple of t. For this situation, Hedayat and Zhao (1990) obtained A-optimal designs as shown in the next two theorems. An A-optimal design is one that estimates the parametric functions of interest, such as direct effect contrasts, with minimum average variance.

Theorem 3.3.4. *Suppose $n = qt + l$, $0 < l < t$. Let $d^* \in \Omega_{t,n,2}$ be a design which allocates only a single treatment to all subjects in period 1 and the allocation in the second period is such that*

$$\begin{aligned} g_{d^*s} &= q + 1, \ \textit{for}\ s = 1, 2, \ldots, l, \\ &= q, \ \textit{for}\ s = l + 1, l + 2, \ldots, t. \end{aligned}$$

Then d^* is the unique A-optimal design for direct effects over $\Omega_{t,n,2}$.

Proof. The block design d_0^* obtained from d^* following the method of Theorem 3.3.1 consists of only a single block given by the second period of d^*. Hence, as one can easily verify, it is A-optimal uniquely since the associated replication numbers are as nearly equal as possible. From (3.3.2), the result is now evident. □

The next example gives one such A-optimal design.

Example 3.3.5. Suppose $t = 6$ and $n = 14$. Then $q = 2, l = 2$ and so, the A-optimal design is as given below.

$$\begin{matrix} 1 & 1 & 1 & 1 & 1 & 1 & 1 & 1 & 1 & 1 & 1 & 1 & 1 & 1 \\ 1 & 1 & 1 & 2 & 2 & 2 & 3 & 3 & 4 & 4 & 5 & 5 & 6 & 6 \end{matrix}. \quad \square$$

Theorem 3.3.5. *Let $d^* \in \Omega_{t,n,2}$ be such that both the periods are identical and f_{d^*s}'s are as equal as possible. Then d^* is A-optimal for carryover effects and is universally optimal if $n = \mu_1 t$, where μ_1 is a positive integer.* □

Further results on the optimality of two-period crossover designs were obtained by Carriere and Reinsel (1993) who assumed the subject effects to be random. We now review some of their results.

Suppose t treatments are to be compared via a two-period crossover design involving n subjects. Clearly, there are $v = t^2$ possible treatment sequences. Carriere and Reinsel (1993) considered the situation when these v treatment sequences are assigned to $n = \sum_{l=1}^{v} n_l$ subjects at random, where n_l is the number of subjects assigned to the treatment sequence l, $1 \leq l \leq v$. Let $d(i,l)$ denote the treatment in the ith position (period) in the treatment sequence l. The model assumed is similar to the one in (3.2.6) and is given by

$$Y_{ijl} = \mu + \alpha_i + \tau_{d(i,l)} + \rho_{d(i-1,l)} + \beta_{jl} + \epsilon_{ijl}, \tag{3.3.4}$$

where Y_{ijl} is the response obtained in period i from the jth subject assigned to the sequence l and μ, α_i, τ_s, ρ_s are as in model (1.2.1); the random subject effects β_{jl} and the errors ϵ_{ijl} are assumed to be mutually uncorrelated random variables with means zero and variances σ_β^2 and σ^2, respectively, $1 \leq i \leq 2$, $1 \leq j \leq n_l$, $1 \leq l \leq v$.

Then for $d \in \Omega_{t,n,2}$, Carriere and Reinsel (1993) noted that the information matrix for the direct effects C_d under model (3.3.4) can be expressed as

$$\sigma^2 C_d = (1+\nu)^{-1}\left\{R_d - \nu^2 \bar{R}_d - n^{-1}(1-\nu^2)\bar{\boldsymbol{r}}_d\bar{\boldsymbol{r}}_d' - Z_d \bar{R}_d^- Z_d'\right\}, \tag{3.3.5}$$

with

$$\nu = \sigma_\beta^2/(\sigma^2 + \sigma_\beta^2)$$

and other notation as in Chapter 1.

Using this, Carriere and Reinsel (1993) proved the following result, invoking Theorem 1.4.1.

Theorem 3.3.6. *Let $d^* \in \Omega_{t,n,2}$ be a strongly balanced design which is uniform on periods. Then under the model (3.3.3), d^* is universally optimal for direct effects over $\Omega_{t,n,2}$.* □

If sequence effects are present in the model, then too, Carriere and Reinsel (1993) showed that the optimality of d^* as in Theorem 3.3.6 is preserved.

Since d^* of Theorem 3.3.6 is strongly balanced, for the existence of such a d^* it is necessary that n is a multiple of t^2. When $n = at^2 + t$ for an integer $a > 0$, Carriere and Reinsel (1993) showed that a design d^* with $z_{d^*ss} = a + 1$ and $z_{d^*ss'} = a$ for $1 \leq s, s' \leq t; s \neq s'$, is universally optimal for direct effects over those designs in $\Omega_{t,n,2}$ which are uniform on the first period. Incidentally, this result establishes the optimality of the design d_2 in Example 3.3.2. Similarly, when $n = at^2 + t(t-1)$, a design d^* with $z_{d^*ss} = a$ and $z_{d^*ss'} = a + 1$ for $1 \leq s, s' \leq t; s \neq s'$, is universally optimal for direct effects over those designs in $\Omega_{t,n,2}$ which are uniform on the first period.

Theorem 3.3.3 shows that under a fixed effects model, some designs which allocate only a single treatment in period 1 can be universally optimal; see design d_1 of Example 3.3.3. This may be contrasted with the findings of Carriere and Reinsel (1993) who observed that these designs no longer remain optimal under the mixed effects model (3.3.4). In fact, for $n = t^2$, they showed that the efficiency of such a design, relative to the optimal design of Theorem 3.3.6, is only $\frac{1}{2-\nu^2}$, in so far as the estimation of direct effects is concerned. Similarly, for $n = at^2 + t$ or $at^2 + t(t-1), a > 0$, this efficiency, relative to the optimal design described in the preceding paragraph, turns out to be $\frac{1}{2-\nu^2-t^2/n^2}$.

3.4 Optimality of Patterson Designs

The balanced crossover designs given by Patterson (1952) for $p \leq t$ are quite popular among experimenters as, for given t, these often involve a moderate number of subjects while keeping p small. These designs have been around for several decades and the efficiencies of these designs are known to be quite high in many cases. Optimality properties of these designs were studied by Shah, Bose and Raghavarao (2005) and in this section, we review some of their findings. In these results, the formulation of the universal optimality criteria is as given by Shah and Sinha (2002) and described in Section 1.4.

All results in this section are under model (1.2.1).

Definition 3.4.1. A design $d \in \Omega_{t,n,p}$ will be said to be a Patterson design if the following conditions hold:

(i) d is uniform on periods, i.e., $n = \mu_1 t$ for some integer μ_1;

(ii) d is balanced, i.e., $Z_d = \lambda_1(J_t - I_t)$ for some integer λ_1;

(iii) when the subjects of d are viewed as blocks, they form the blocks of a balanced incomplete block (BIB) design with block size p ;

(iv) when d is restricted to the first $p-1$ periods, then again, the subjects of d form the blocks of a BIB design with block size $p-1$;

(v) in the set of μ_1 subjects receiving a given treatment in the last period, every other treatment is applied λ_1 times in the first $p-1$ periods.

The conditions (i)–(v) above are equivalent to the conditions given by Patterson (1952). It may be noted that crossover designs satisfying these conditions were called "balanced" by Patterson (1952). However, subsequently, the term "balance" has been popularly used in the sense of Definition 2.2.4 and so, in order to avoid confusion, we shall use the term *Patterson designs* for designs satisfying the conditions of Definition 3.4.1.

From conditions (i) and (ii) of Definition 3.4.1 it follows as in Section 1.2 that the parameters of a Patterson design must necessarily satisfy

$$p \leq t, \quad n = \mu_1 t \text{ and } \lambda_1 = \mu_1(p-1)/(t-1).$$

In what follows, we exclude the case $p = t = 2, n = 2\mu_1$, even though designs satisfying the conditions (i)–(v) exist for these parametric values; this is simply because in such a case, neither the contrasts among direct effects nor those among the carryover effects are estimable. Recalling from Section 1.3 the notion of connectedness of a crossover design, it can be seen that the Patterson designs are connected for other values of t, n and p.

Example 3.4.1. A Patterson design with $t = 4, p = 3, n = 12$ is shown below.

$$\begin{array}{cccccccccccc} 1 & 2 & 3 & 4 & 1 & 2 & 3 & 4 & 1 & 2 & 3 & 4 \\ 2 & 1 & 4 & 3 & 3 & 4 & 1 & 2 & 4 & 3 & 2 & 1 \\ 3 & 4 & 1 & 2 & 4 & 3 & 2 & 1 & 2 & 1 & 4 & 3 \end{array}. \qquad \square$$

Let $\mathcal{B}_{t,n,p}$ consist of connected designs in $\Omega_{t,n,p}$ that are binary on subjects in the sense that a treatment is applied to a subject at most once. Also, for $d \in \Omega_{t,n,p}$, let $C_d(\boldsymbol{\tau}, \boldsymbol{\rho}, \boldsymbol{\alpha})$ denote the information matrix for estimating $(\boldsymbol{\tau}, \boldsymbol{\rho}, \boldsymbol{\alpha})$. Using the notation introduced in Chapter 1, under the model (1.2.1) it follows from (1.3.7) that for $d \in \Omega_{t,n,p}$, this symmetric matrix is given by

$$C_d(\boldsymbol{\tau}, \boldsymbol{\rho}, \boldsymbol{\alpha}) = \begin{bmatrix} R_d - p^{-1}N_dN_d' & Z_d - p^{-1}N_d\bar{N}_d' & M_d - p^{-1}N_dJ_{np} \\ & \bar{R}_d - p^{-1}\bar{N}_d\bar{N}_d' & \bar{M}_d - p^{-1}\bar{N}_dJ_{np} \\ & & nI_p - np^{-1}J_p \end{bmatrix}. \tag{3.4.1}$$

The information matrix for joint estimation of $(\boldsymbol{\tau}, \boldsymbol{\rho})$ is as usual denoted by $C_d(\boldsymbol{\tau}, \boldsymbol{\rho})$ as in (1.3.9).

Let $d^* \in \Omega_{t,n,p}$ be a Patterson design. Then from Definition 3.4.1 it follows that

$$\begin{aligned} R_{d^*} &= \mu_1 pI_t, \quad \bar{R}_{d^*} = \mu_1(p-1)I_t, \\ M_{d^*} &= \mu_1 J_{tp}, \quad \bar{M}_{d^*} = [\mathbf{0}_t \ \ \mu_1 J_{t,p-1}], \\ Z_{d^*} &= \lambda_1(J_t - I_t), \\ N_{d^*}N_{d^*}' &= p(\mu_1 - \lambda_1)I_t + p\lambda_1 J_t, \\ \bar{N}_{d^*}\bar{N}_{d^*}' &= ((p-1)\mu_1 - (p-2)\lambda_1)I_t + (p-2)\lambda_1 J_t. \end{aligned} \tag{3.4.2}$$

The first two lines in (3.4.2) follow as in Lemma 2.2.1 from condition (i) of Definition 3.4.1, the third line is due to condition (ii) and the remaining two lines are consequences of conditions (iii) and (iv), respectively.

For any design $d \in \Omega_{t,n,p}$, without loss of generality, we can rearrange the subjects so that the first n_1 subjects have treatment 1 in the last period, the next n_2 subjects have treatment 2 in the last period, ..., the last n_t subjects have treatment t in the last period, where $\sum_{s=1}^{t} n_s = n$. With this arrangement of the subjects, it is clear that

$$\bar{N}_d' = N_d' - \begin{bmatrix} \mathbf{1}_{n_1} & \mathbf{0} & \cdots & \mathbf{0} \\ \mathbf{0} & \mathbf{1}_{n_2} & \cdots & \mathbf{0} \\ \vdots & & & \\ \mathbf{0} & \mathbf{0} & \cdots & \mathbf{1}_{n_t} \end{bmatrix}.$$

It can now be seen that

$$N_d\bar{N}_d' = N_dN_d' - \Theta_d, \quad \text{and } \bar{N}_d\bar{N}_d' = N_dN_d' - \Theta_d - \Theta_d' + \operatorname{diag}(n_1, \ldots, n_t),$$

where $\Theta_d = (\boldsymbol{\theta}_1, \ldots, \boldsymbol{\theta}_t)$ and for $1 \leq s \leq t$, $\boldsymbol{\theta}_s$ is the $t \times 1$ vector obtained by taking the sum of the n_s columns of N_d corresponding to the n_s subjects with treatment s in the last period.

Again, from Definition 3.4.1 it follows that for a Patterson design d^*, we have

$$\begin{aligned} \Theta_{d^*} &= (\mu_1 - \lambda_1)I_t + \lambda_1 J_t, \\ N_{d^*}N'_{d^*} &= p\Theta_{d^*}, \\ N_{d^*}\bar{N}'_{d^*} &= (p-1)\Theta_{d^*}, \\ \bar{N}_{d^*}\bar{N}'_{d^*} &= (p-2)\Theta_{d^*} + \mu_1 I_t. \end{aligned} \tag{3.4.3}$$

From (3.4.1), on using (3.4.2) and (3.4.3) we find that $C_{d^*}(\boldsymbol{\tau}, \boldsymbol{\rho}, \boldsymbol{\alpha})$ is equal to

$$\begin{bmatrix} \frac{t\mu_1(p-1)}{(t-1)}H_t & -\frac{t\mu_1(p-1)}{p(t-1)}H_t & \mathbf{0}_{tp} \\ -\frac{t\mu_1(p-1)}{p(t-1)}H_t & \frac{\mu_1(p-1)(pt-t-1)}{p(t-1)}H_t + \frac{\mu_1(p-1)}{pt}J_t & \frac{\mu_1}{p}\mathbf{1}_t\left(-(p-1), \mathbf{1}'_{p-1}\right) \\ \mathbf{0}_{pt} & \frac{\mu_1}{p}\begin{pmatrix} -(p-1) \\ \mathbf{1}_{p-1} \end{pmatrix}\mathbf{1}'_t & nI_p - \frac{n}{p}J_p \end{bmatrix}, \tag{3.4.4}$$

where, as before, $H_t = I_t - t^{-1}J_t$.

From (3.4.4), we can write $C_{d^*}(\boldsymbol{\tau}, \boldsymbol{\rho})$ as in (1.3.9) with

$$\begin{aligned} C_{d^*11} &= \frac{\mu_1 t(p-1)}{(t-1)}H_t, \\ C_{d^*12} &= -\frac{t\mu_1(p-1)}{p(t-1)}H_t = C'_{d^*21}, \\ C_{d^*22} &= \frac{\mu_1(p-1)(pt-t-1)}{p(t-1)}H_t. \end{aligned} \tag{3.4.5}$$

To prove the next result we need the following lemma, which can be proved easily.

Lemma 3.4.1. *Let $A = (a_{uv})$ be a $t \times t$ matrix and let g be a permutation on $\{1, 2, \ldots, t\}$, the corresponding permutation matrix being denoted by E_g. Let S_t be the symmetric group on $\{1, 2, \ldots, t\}$. Then* $\bar{A} = \frac{1}{t!}\sum_{g \in S_t} E'_g A E_g$

is a completely symmetric matrix with diagonal elements equal to a and off-diagonal elements equal to b, where

$$a = \frac{1}{t}\sum_{u=1}^{t} a_{uu}, \quad b = \frac{1}{t(t-1)}\left(\sum_{u=1}^{t}\sum_{v=1}^{t} a_{uv} - \sum_{u=1}^{t} a_{uu}\right).$$

□

For an arbitrary design d, let d_g denote the design obtained from it by permuting the t treatment labels according to g, $g \in S_t$. We then have the following result.

Lemma 3.4.2. *For a Patterson design $d^* \in \Omega_{t,n,p}$,*

$$C_{d^*}(\boldsymbol{\tau}, \boldsymbol{\rho}, \boldsymbol{\alpha}) = \frac{1}{t!}\sum_{g \in S_t} C_{d_g}(\boldsymbol{\tau}, \boldsymbol{\rho}, \boldsymbol{\alpha}), \tag{3.4.6}$$

where $d \in \mathcal{B}_{t,n,p}$.

Proof. It is enough to consider the different submatrices of $C_d(\boldsymbol{\tau}, \boldsymbol{\rho}, \boldsymbol{\alpha})$ in (3.4.1) for a design $d \in \mathcal{B}_{t,n,p}$ and show that for each of these, the average over all permutations of treatment labels equals the corresponding expression in $C_{d^*}(\boldsymbol{\tau}, \boldsymbol{\rho}, \boldsymbol{\alpha})$ as given in (3.4.4).

Note that $C_{d_g}(\boldsymbol{\tau}, \boldsymbol{\rho}, \boldsymbol{\alpha}) = Q_g' C_d(\boldsymbol{\tau}, \boldsymbol{\rho}, \boldsymbol{\alpha}) Q_g$, where

$$Q_g = \begin{pmatrix} E_g & \mathbf{0}_{tt} & \mathbf{0}_{tt} \\ \mathbf{0}_{tt} & E_g & \mathbf{0}_{tt} \\ \mathbf{0}_{tt} & \mathbf{0}_{tt} & I_t \end{pmatrix}$$

and E_g is as defined in Lemma 3.4.1. Using Lemma 3.4.1, from (3.4.1), it can be easily checked that for a connected design $d \in \mathcal{B}_{t,n,p}$, the averaged versions of each of $R_d - p^{-1}N_dN_d'$, $Z_d - p^{-1}N_d\bar{N}_d'$ and $\bar{R}_d - p^{-1}\bar{N}_d\bar{N}_d'$ is equal to the corresponding expression for d^* as shown in (3.4.4). For example, since d is binary over subjects, the sth diagonal element of N_dN_d' is $\sum_j n_{dsj}^2 = \sum_j n_{dsj} = r_{ds}$ and the average of r_{ds} over all permutations is $p\mu_1 (= r_{d^*s})$ for all s. Hence, since $\mathbf{1_t}'N_dN_d'\mathbf{1}_t = p^2 n$, applying Lemma 3.4.1, it can be seen that the average of $R_d - p^{-1}N_dN_d'$ is equal to $\dfrac{t\mu_1(p-1)}{t-1}H_t$ which is equal to $R_{d^*} - p^{-1}N_{d^*}N_{d^*}'$, as shown in (3.4.4). Moreover, for a binary design d, the sth diagonal element of Θ_d is n_s. The average of n_s over all permutations is $\mu_1 = n/t$ and this is equal to the common diagonal element of Θ_{d^*}. It can also be seen that the averages of M_d and N_d are $\mu_1 J_{tp}$ and $pt^{-1}J_{tn}$, respectively. Thus the average of $M_d - p^{-1}N_dJ_{np}$ is $\mathbf{0}_{tp}$, which equals the corresponding term for d^* as given in (3.4.4). Similarly, the average of $\bar{M}_d - p^{-1}\bar{N}_d J_{np}$ leads to the corresponding expression in (3.4.4). This completes the proof. □

In the following theorem, we prove a strong optimality property for d^*.

Theorem 3.4.1. *Let $d^* \in \Omega_{t,n,p}$ be a Patterson design. Then d^* is universally optimal for the joint estimation of direct and carryover effects over $\mathcal{B}_{t,n,p}$.*

Proof. Obtaining the information matrix $C_d(\boldsymbol{\tau}, \boldsymbol{\rho})$ from $C_d(\boldsymbol{\tau}, \boldsymbol{\rho}, \boldsymbol{\alpha})$ by adjusting for the period effects $\boldsymbol{\alpha}$ is equivalent to computing the Schur complement of the diagonal block corresponding to $\boldsymbol{\alpha}$ in $C_d(\boldsymbol{\tau}, \boldsymbol{\rho}, \boldsymbol{\alpha})$. This Schur complement in $C_{d^*}(\boldsymbol{\tau}, \boldsymbol{\rho}, \boldsymbol{\alpha})$ is $C_{d^*}(\boldsymbol{\tau}, \boldsymbol{\rho})$ and that in $C_{dg}(\boldsymbol{\tau}, \boldsymbol{\rho}, \boldsymbol{\alpha})$ is $C_{dg}(\boldsymbol{\tau}, \boldsymbol{\rho})$. Since the Schur complement is a concave function (see e.g., Pukelsheim (1993)), from (3.4.6) it is evident that

$$C_{d^*}(\boldsymbol{\tau}, \boldsymbol{\rho}) \geq \frac{1}{t!} \sum_{g \in S_t} C_{dg}(\boldsymbol{\tau}, \boldsymbol{\rho}).$$

Using the sufficient condition (1.4.1) with weights $w_g = 1/t!$, $g \in S_t$, it follows that d^* is universally optimal for the joint estimation of $(\boldsymbol{\tau}, \boldsymbol{\rho})$ over $\mathcal{B}_{t,n,p}$. □

Remark 3.4.1. Markiewicz (1997) and Shah and Sinha (2002) showed that universal optimality for the joint estimation of two sets of parameters implies the universal optimality for the estimation of each set of parameter (though the converse is not true) and much more. So, Theorem 3.4.1 implies that d^* is universally optimal for the estimation of direct as well as carryover effects over $\mathcal{B}_{t,n,p}$. □

We next consider the optimality properties of d^* over a wider class than $\mathcal{B}_{t,n,p}$. Let $\Lambda_{t,n,p}$ denote the class of connected designs in $\Omega_{t,n,p}$ in which no treatment is assigned to two consecutive periods on the same subject. Clearly, $\mathcal{B}_{t,n,p} \subset \Lambda_{t,n,p} \subset \Omega_{t,n,p}$.

To simplify matters, we initially assume a simpler model than (1.3.3) in which we assume that there are no period effects, all other terms being as in (1.3.3); this assumption is relaxed later. Under such a model, for a design $d \in \Lambda_{t,n,p}$, let $\tilde{C}_d(\boldsymbol{\tau}, \boldsymbol{\rho})$ denote the information matrix for estimating $\boldsymbol{\tau}$ and $\boldsymbol{\rho}$ jointly, and $\tilde{C}_d$ be the information matrix for estimating direct effects. Then the submatrices of $\tilde{C}_d(\boldsymbol{\tau}, \boldsymbol{\rho})$ are

$$\begin{aligned} \tilde{C}_{d11} &= R_d - p^{-1}N_d N_d', \\ \tilde{C}_{d12} &= Z_d - p^{-1}N_d \bar{N}_d' = \tilde{C}_{d21}', \\ \tilde{C}_{d22} &= \bar{R}_d - p^{-1}\bar{N}_d \bar{N}_d'. \end{aligned} \tag{3.4.7}$$

The corresponding expressions for a Patterson design d^* are given by

$$\begin{aligned}
\tilde{C}_{d^*11} &= \frac{t\mu_1(p-1)}{t-1}H_t, \\
\tilde{C}_{d^*12} &= -\frac{t\mu_1(p-1)}{p(t-1)}H_t = \tilde{C}'_{d^*21}, \\
\tilde{C}_{d^*22} &= \frac{\mu_1(p-1)(pt-t-1)}{p(t-1)}H_t + \frac{\mu_1(p-1)}{pt}J_t.
\end{aligned} \tag{3.4.8}$$

For $d \in \Lambda_{t,n,p}$, let $A_d(\boldsymbol{\tau}, \boldsymbol{\rho})$ denote the average of $\tilde{C}_d(\boldsymbol{\tau}, \boldsymbol{\rho})$, the average being over all permutation of treatment labels, i.e.,

$$A_d(\boldsymbol{\tau}, \boldsymbol{\rho}) = \frac{1}{t!}\sum_{g \in S_t} \tilde{C}_{d_g}(\boldsymbol{\tau}, \boldsymbol{\rho}) = \begin{pmatrix} A_{d11} & A_{d12} \\ A_{d21} & A_{d22} \end{pmatrix}, \text{ (say).} \tag{3.4.9}$$

From (3.4.7), arguments similar to those used in proving Lemma 3.4.2 yield

$$\begin{aligned}
A_{d11} &= \frac{p^2t\mu_1 - \beta}{p(t-1)}H_t, \\
A_{d12} &= -\frac{\beta - l}{p(t-1)}H_t, \\
A_{d22} &= \frac{pt\mu_1(p-1) - (\beta - 2l) - \mu_1(t+p-1)}{p(t-1)}H_t \\
&\quad + \frac{\mu_1(p-1)}{pt}J_t,
\end{aligned} \tag{3.4.10}$$

where

$$\begin{aligned}
\beta &= \sum_s \sum_j n_{dsj}^2, \\
l &= \sum_s \text{ (sum of } n_{dsj} \text{ for subjects with treatment } s \text{ in the last period).}
\end{aligned} \tag{3.4.11}$$

The following theorem shows the optimality property of a Patterson design over the class $\Lambda_{t,n,p}$. We only give an outline of its proof below and refer to Shah *et al.* (2005) for details.

Theorem 3.4.2. *Let $d^* \in \Omega_{t,n,p}$ be a Patterson design. Then d^* is universally optimal for direct as well as carryover effects over $\Lambda_{t,n,p}$.*

Proof. Let d be any design in $\Lambda_{t,n,p}$. Shah *et al.* (2005) showed that $\tilde{C}_{d^*22}$ and A_{d22} are both nonsingular and hence, $\tilde{C}_{d^*22}^{-1}$ and A_{d22}^{-1} exist. Then as in (1.3.11), with $\tilde{C}_{d^*} = \tilde{C}_{d^*11} - \tilde{C}_{d^*12}\tilde{C}_{d^*22}^{-1}\tilde{C}_{d^*21}$ and similarly, writing $A_d = A_{d11} - A_{d12}A_{d22}^{-1}A_{d21}$, it follows that

$$\tilde{C}_{d^*} = \frac{t\mu_1(p-1)}{t-1}\left[1 - \frac{t}{p(pt-t-1)}\right] H_t,$$

$$A_d = \frac{1}{p(t-1)}\left[p^2 t\mu_1 - \beta - \frac{(\beta-l)^2}{pt\mu_1(p-1) - (\beta-2l) - \mu_1(t+p-1)}\right] H_t, \tag{3.4.12}$$

where β and l are as in (3.4.11).

At this stage, some more notation will be helpful. For any symmetric matrix

$$B = \begin{pmatrix} B_{11} & B_{12} \\ B_{21} & B_{22} \end{pmatrix},$$

where each B_{uv} is $t \times t$, let

$$B_{11.22} = B_{11} - B_{12}B_{22}^{-}B_{21}.$$

Then the information matrix $\tilde{C}_{d^*}$ and the matrix A_d shown in (3.4.12) can be equivalently written as

$$\tilde{C}_{d^*} = \tilde{C}_{d^*}(\boldsymbol{\tau}, \boldsymbol{\rho})_{11.22} \text{ and } A_d = A_d(\boldsymbol{\tau}, \boldsymbol{\rho})_{11.22}$$

where the submatrices comprising $\tilde{C}_{d^*}(\boldsymbol{\tau}, \boldsymbol{\rho})$, and $A_d(\boldsymbol{\tau}, \boldsymbol{\rho})$ are as given in (3.4.8) and (3.4.10), respectively.

We first show that d^* is universally optimal for direct effects over $\Lambda_{t,n,p}$ when period effects are assumed absent. For this, by (1.4.1), it is enough to show that

$$\tilde{C}_{d^*}(\boldsymbol{\tau}, \boldsymbol{\rho})_{11.22} \geq \frac{1}{t!}\sum_{g \in S_t} \tilde{C}_{dg}(\boldsymbol{\tau}, \boldsymbol{\rho})_{11.22}. \tag{3.4.13}$$

Shah *et al.* (2005) showed that

$$\tilde{C}_{d^*}(\boldsymbol{\tau}, \boldsymbol{\rho})_{11.22} \geq A_d(\boldsymbol{\tau}, \boldsymbol{\rho})_{11.22} \tag{3.4.14}$$

and it is easy to see that

$$\tilde{C}_{dg}(\boldsymbol{\tau}, \boldsymbol{\rho})_{11.22} = \left(\tilde{C}_d(\boldsymbol{\tau}, \boldsymbol{\rho})_{11.22}\right)_g,$$

where the right-hand side is obtained by applying g to the rows and columns of $\tilde{C}_d(\boldsymbol{\tau}, \boldsymbol{\rho})_{11.22}$.

Recalling that the Schur complement is a concave function, we have

$$\left(\sum_{g \in S_t} \tilde{C}_{dg}(\boldsymbol{\tau}, \boldsymbol{\rho})\right)_{11.22} \geq \sum_{g \in S_t} \tilde{C}_{dg}(\boldsymbol{\tau}, \boldsymbol{\rho})_{11.22}. \tag{3.4.15}$$

Therefore, by (3.4.14), (3.4.9) and (3.4.15),

$$\begin{aligned}\tilde{C}_{d^*}(\boldsymbol{\tau},\boldsymbol{\rho})_{11.22} \geq A_d(\boldsymbol{\tau},\boldsymbol{\rho})_{11.22} &= \frac{1}{t!}\left(\sum_{g\in S_t}\tilde{C}_{dg}(\boldsymbol{\tau},\boldsymbol{\rho})\right)_{11.22} \\ &\geq \frac{1}{t!}\sum_{g\in S_t}\tilde{C}_{dg}(\boldsymbol{\tau},\boldsymbol{\rho})_{11.22}.\end{aligned} \tag{3.4.16}$$

Hence (3.4.13) holds and d^* is universally optimal over $\Lambda_{t,n,p}$ for a model without the period effects.

Now, under model (1.2.1) (i.e., when period effects are present), remembering that C_d, the information matrix for direct effects under this model, is equivalently written as $C_d = C_d(\boldsymbol{\tau},\boldsymbol{\rho})_{11.22}$, it is easy to see as in (1.4.6) that

$$C_d(\boldsymbol{\tau},\boldsymbol{\rho})_{11.22} \leq \tilde{C}_d(\boldsymbol{\tau},\boldsymbol{\rho})_{11.22},$$

with equality holding if d is uniform on periods. Hence, using (3.4.16),

$$\begin{aligned}C_{d^*}(\boldsymbol{\tau},\boldsymbol{\rho})_{11.22} = \tilde{C}_{d^*}(\boldsymbol{\tau},\boldsymbol{\rho})_{11.22} &\geq \frac{1}{t!}\sum_{g\in S_t}\tilde{C}_{dg}(\boldsymbol{\tau},\boldsymbol{\rho})_{11.22} \\ &\geq \frac{1}{t!}\sum_{g\in S_t}C_{dg}(\boldsymbol{\tau},\boldsymbol{\rho})_{11.22}\end{aligned}$$

for every $d \in \Lambda_{t,n,p}$. By (1.4.1), this establishes the universal optimality of d^* for direct effects under model (1.2.1). An analogous treatment will show the universal optimality of d^* for carryover effects where one has to work with a g-inverse of C_{d11}. □

Remark 3.4.2. The design d^* is *not* universally optimal over $\Lambda_{t,n,p}$ for the joint estimation of the direct and carryover effects (*cf.* Shah and Sinha (2002)). It is also not known whether a Patterson design is universally optimal for the estimation of direct or carryover effects over the entire class $\Omega_{t,n,p}$. However, as demonstrated later in Section 4.8 (see Table 4.8.1), many Patterson designs have very high efficiencies for the separate estimation of both direct and carryover effects over $\Omega_{t,n,p}$. □

3.5 Constructions

Patterson (1952) provided construction methods of several families of Patterson designs with $p \leq t$. In what follows, we describe some of these constructions.

The first method is applicable to the case where t is a prime or a prime power, $p \leq t$ and $n = t(t-1)$. This method uses a complete set of mutually orthogonal Latin squares (MOLS) of order t. Recall that a Latin square of order $m(\geq 2)$ is an $m \times m$ array, with entries from a set of m distinct symbols such that each symbol appears exactly once in each row and once in each column of the array. Two Latin squares of the same order are said to be orthogonal to each other if, when any one of the squares is superimposed on the other, every ordered pair of symbols appears exactly once. A set of Latin squares is said to form a set of MOLS if every pair in the set is orthogonal to each other. The maximum number of MOLS of order $m(> 2)$ is $(m-1)$, this number being attainable if m is a prime or a prime power (see e.g., Raghavarao (1971, Chapter 1)) and in such a case we say that there is a complete set of MOLS.

We now have the following result.

Theorem 3.5.1. *For t a prime or a prime power, a Patterson design exists in $\Omega_{t,n=t(t-1),p}$, where $p \leq t$.*

Proof. Let t be a prime or a prime power and consider a complete set of $t-1$ MOLS of order t, such that any square in the set is obtained from any other square of the set by permuting the rows. Such a set of MOLS can be constructed as follows (see e.g., Stevens (1939)). Let x be a primitive element of $GF(t)$ and let the elements of $GF(t)$ be represented by $u_0 = 0, u_1 = 1, u_2, \ldots, u_{t-1}$. For $0 \leq i \leq t-2$, consider the arrays

$$L_i = \begin{array}{cccc} 0 & 1 & \cdots & u_{t-1} \\ x^{0+i} & x^{0+i}+1 & \cdots & x^{0+i}+u_{t-1} \\ x^{1+i} & x^{1+i}+1 & \cdots & x^{1+i}+u_{t-1} \\ \vdots & & & \\ x^{t-2+i} & x^{t-2+i}+1 & \cdots & x^{t-2+i}+u_{t-1} \end{array} .$$

Then the set $\{L_i\}$, $0 \leq i \leq t-2$, forms a complete set of MOLS of order t where L_{i+1} can be obtained by cyclically permuting the last $t-1$ rows of L_i. Form the $t \times t(t-1)$ array L given by

$$L = [L_0 \; L_1 \; \cdots \; L_{t-2}].$$

This L gives a Patterson design in $\Omega_{t,t(t-1),t}$, where the rows of L represent the periods and columns, the subjects. Now, on deleting any $t-p$ rows of L, one gets a design in $\Omega_{t,t(t-1),p}$ with $p < t$. Patterson (1952) showed that this design satisfies the conditions of a Patterson design. □

Example 3.5.1. Let $t = 4$. For notational simplicity, we use the mapping

$$u_0 \to 1, \quad u_1 \to 2, \quad u_2 \to 3, \quad u_3 \to 4.$$

Then the three MOLS of order 4, constructed by the method described above, are:

$$L_0 = \begin{bmatrix} 1\,2\,3\,4 \\ 2\,1\,4\,3 \\ 3\,4\,1\,2 \\ 4\,3\,2\,1 \end{bmatrix}, \ L_1 = \begin{bmatrix} 1\,2\,3\,4 \\ 3\,4\,1\,2 \\ 4\,3\,2\,1 \\ 2\,1\,4\,3 \end{bmatrix}, \ L_2 = \begin{bmatrix} 1\,2\,3\,4 \\ 4\,3\,2\,1 \\ 2\,1\,4\,3 \\ 3\,4\,1\,2 \end{bmatrix}.$$

Now deleting the last row of $L = [L_0\ L_1\ L_2]$, one gets the Patterson design exhibited in Example 3.4.1. □

Example 3.5.2. Let $t = 5$. Here,

$$L = \begin{bmatrix} 0\,1\,2\,4\,3 & 0\,1\,2\,4\,3 & 0\,1\,2\,4\,3 & 0\,1\,2\,4\,3 \\ 1\,2\,3\,0\,4 & 2\,3\,4\,1\,0 & 4\,0\,1\,3\,2 & 3\,4\,0\,2\,1 \\ 2\,3\,4\,1\,0 & 4\,0\,1\,3\,2 & 3\,4\,0\,2\,1 & 1\,2\,3\,0\,4 \\ 4\,0\,1\,3\,2 & 3\,4\,0\,2\,1 & 1\,2\,3\,0\,4 & 2\,3\,4\,1\,0 \\ 3\,4\,0\,2\,1 & 1\,2\,3\,0\,4 & 2\,3\,4\,1\,0 & 4\,0\,1\,3\,2 \end{bmatrix}.$$

If a design with $p = 3$ is required, simply deleting the last two rows of L gives a Patterson design in $\Omega_{5,20,3}$. □

Remark 3.5.1. Note that the result of Theorem 3.5.1 does *not* extend to all sets of MOLS. For instance, the method of construction of a complete set of MOLS given by Bose and Nair (1941) does *not* have the property that each Latin square in the set can be obtained by a permutation of rows of other squares in the set and thus, these MOLS cannot be used in the construction of Theorem 3.5.1. □

A disadvantage of the method of construction in Theorem 3.5.1 is that the design requires $n = t(t-1)$ and this number can be quite large for even moderate values of t. For smaller values of n, Patterson (1952) constructed designs using the method of differences. Let the treatments be represented by the elements of an additive Abelian group of order t, with 0 denoting the identity of the group. In this method, it suffices to find n/t initial sequences, each having p elements, satisfying certain properties. The design in $\Omega_{t,n,p}$ is obtained by "developing" each of these initial sequences, each initial sequence giving rise to a set of t sequences. Here, by the term "developing" we mean that other sequences of a set are obtained by adding the non-zero elements of the group in turn to each element in the initial sequence.

The initial sequences needed for the construction of Patterson designs must satisfy certain conditions. Without loss of generality, the first treatment in each of the initial sequences can be taken to be 0, the identity of the group. Clearly, a treatment must occur at most once in each initial sequence. Suppose the differences between successive elements of the kth initial sequence are $d_{1k}, d_{2k}, \ldots, d_{p-1,k}$, $1 \le k \le n/t$.

For $1 \le k \le n/t$, define

$$\delta_{1k} = d_{1k},\ \delta_{2k} = d_{1k} + d_{2k},\ \ldots, \delta_{p-1,k} = d_{1k} + d_{2k} + \cdots + d_{p-1,k}.$$

Thus, the initial sequence is $(0, \delta_{1k}, \delta_{2k}, \ldots, \delta_{p-1,k})'$. Now, consider the arrays

$$\begin{array}{ccccc} 0 & -\delta_{1k} & -\delta_{2k} & \cdots & -\delta_{p-1,k} \\ \delta_{1k} & 0 & -(\delta_{2k} - \delta_{1k}) & \cdots & -(\delta_{p-1,k} - \delta_{1,k}) \\ \delta_{2k} & (\delta_{2k} - \delta_{1k}) & 0 & \cdots & -(\delta_{p-1,k} - \delta_{2,k}) \\ \vdots & & & & \vdots \\ \delta_{p-1,k} & (\delta_{p-1,k} - \delta_{1k}) & (\delta_{p-1,k} - \delta_{2,k}) & \cdots & 0 \end{array}. \quad (3.5.1)$$

For $1 \le k \le n/t$, let $A_{1k} = \{d_{1k}, d_{2k}, \ldots, d_{p-1,k}\}$ and $A_1 = \bigcup_{k=1}^{n/t} A_{1k}$. Also, let A_{2k}, $1 \le k \le n/t$, denote the set of elements to the left of the principal diagonal in the array (3.5.1) and $A_2 = \bigcup_{k=1}^{n/t} A_{2k}$. Similarly, let $A_{3k} = \{d_{p-1,k}, d_{p-2,k} + d_{p-1,k}, \ldots, d_{1k} + d_{2k} + \cdots + d_{p-1,k}\}, 1 \le k \le n/t$ and $A_3 = \bigcup_{k=1}^{n/t} A_{3k}$. Note that the set unions A_1, A_2 and A_3 allow the inclusion of the same element more than once. All the conditions (i)–(v) required for a Patterson design are met if the following hold:

(a) A_2 and $-A_2$ together include all the non-zero elements of the group an equal number of times,

(b) each of A_1 and A_3 includes each non-zero element of the group equally often.

Example 3.5.3. Let $t = 7, p = 3$. Consider the Abelian group of residues mod 7. We start with the three initial sequences

$$(0,5,6)',\ (0,3,5)',\ (0,6,3)'.$$

It is easy to check that

$$A_1 = \{5,1,3,2,6,4\},$$
$$A_2 = \{5,1,6,3,2,5,6,4,3\},$$
$$A_3 = \{1,6,2,5,4,3\}.$$

Clearly, the sets A_i, $1 \le i \le 3$, satisfy the conditions (a) and (b) above and by developing these initial sequences, one gets a Patterson design with $t = 7, n = 21, p = 3$. □

Based on the above idea, Patterson (1952) obtained several families of Patterson designs through the method of differences, some of which are described next.

Theorem 3.5.2. *Let $p = 3$ and suppose t is a prime or a prime power of the form $4m + 3$, where m is a positive integer. Then a Patterson design exists in $\Omega_{t,n=t(t-1)/2,p=3}$.*

Proof. In keeping with the preceding discussion, here we need $k = n/t = (t-1)/2$ initial sequences, such that on developing these we get the treatment sequences for the $n = tk = t(t-1)/2$ subjects. Since $p = 3$, the kth initial sequence will give rise to two differences, d_{1k} and d_{2k}, $1 \leq k \leq (t-1)/2$. For a given k, suppose we write the difference pairs as $\begin{pmatrix} d_{1k} \\ d_{2k} \end{pmatrix}$.

Let x be a primitive element of $GF(t)$. If $x + 1$ is an even power of x, consider the following k difference pairs

$$\begin{pmatrix} x^0 \\ x^1 \end{pmatrix}, \begin{pmatrix} x^2 \\ x^3 \end{pmatrix}, \begin{pmatrix} x^4 \\ x^5 \end{pmatrix}, \ldots, \begin{pmatrix} x^{t-3} \\ x^{t-2} \end{pmatrix}, \tag{3.5.2}$$

and if $x + 1$ is an odd power of x, consider the k difference pairs

$$\begin{pmatrix} x^1 \\ x^0 \end{pmatrix}, \begin{pmatrix} x^3 \\ x^2 \end{pmatrix}, \begin{pmatrix} x^5 \\ x^4 \end{pmatrix}, \ldots, \begin{pmatrix} x^{t-2} \\ x^{t-3} \end{pmatrix}. \tag{3.5.3}$$

Since the first treatment in each of the k initial sequences can be taken to be 0, the initial sequences can now be obtained easily from the difference pairs in (3.5.2) and (3.5.3).

It can be verified that in either case, these initial sequences derived from the sets of differences displayed above, satisfy the conditions (a) and (b). Therefore, using either of (3.5.2) or (3.5.3), a Patterson design in $\Omega_{t=4m+3,n=t(t-1)/2,p=3}$ can be constructed for a positive integer m. □

Example 3.5.4. Let $t = 11$. Then $m = 2, n = 55$ and $k = 5$. A primitive element of $GF(11)$ is $x = 2$ and thus, $x + 1 = 3 = x^8$. Therefore, using (3.5.2), we get the following 5 pairs of differences:

$$\begin{pmatrix} 1 \\ 2 \end{pmatrix}, \begin{pmatrix} 4 \\ 8 \end{pmatrix}, \begin{pmatrix} 5 \\ 10 \end{pmatrix}, \begin{pmatrix} 9 \\ 7 \end{pmatrix}, \begin{pmatrix} 3 \\ 6 \end{pmatrix}.$$

The five initial sequences are thus $(0, 1, 3)', (0, 4, 1)', (0, 5, 4)', (0, 9, 5)', (0, 3, 9)'$. Developing each of these initial sequences mod 11, we get a Patterson design with $t = 11, n = 55, p = 3$. □

The above method of construction has been generalized by Patterson (1952) to obtain some Patterson designs with odd number of periods. For example, a design with $t = 7, n = 21, p = 5$ can be obtained by developing each of the following three initial sequences mod 7:

$$(0,5,6,2,4)' \ (0,3,5,4,1)' \ (0,6,3,1,2)'.$$

For more constructions of designs of this type, we refer the reader to Patterson (1952).

Next, let $t = 4u+3$ be a prime number, where u is a positive integer, and let the treatments be represented by the elements of the group of residues mod t. Patterson (1952) proved the following result.

Theorem 3.5.3. *Suppose t is a prime of the form $4u+3$, where u is a positive integer. Then a Patterson design exists in $\Omega_{t,n=2t,p=(t+1)/2}$.*

Proof. Since $n/t = 2$, we have $k = 2$ and we need two initial sequences. Consider the two sets of differences

$$\{2w, 4w, 6w, \ldots, (t-1)w\} \text{ and } \{-2w, -4w, -6w, \ldots, -(t-1)w\}, \quad (3.5.4)$$

where w is a positive integer satisfying $w \neq 0$ mod t and the differences in (3.5.4) are reduced mod t. Patterson (1952) showed that the initial sequences derived from the sets of differences (3.5.4) satisfy the conditions (a) and (b). □

Example 3.5.5. Let $t = 11$. Choosing $w = 2$, we have, from (3.5.4) the two sets of differences as $\{4,8,1,5,9\}$ and $\{7,3,10,6,2\}$. The initial sequences are thus $(0,4,1,2,7,5)'$ and $(0,7,10,9,4,6)'$. The required Patterson design with $t = 11, n = 22, p = 6$ is shown next.

0	1	2	3	4	5	6	7	8	9	10	0	1	2	3	4	5	6	7	8	9	10
4	5	6	7	8	9	10	0	1	2	3	7	8	9	10	0	1	2	3	4	5	6
1	2	3	4	5	6	7	8	9	10	0	10	0	1	2	3	4	5	6	7	8	9
2	3	4	5	6	7	8	9	10	0	1	9	10	0	1	2	3	4	5	6	7	8.
7	8	9	10	0	1	2	3	4	5	6	4	5	6	7	8	9	10	0	1	2	3
5	6	7	8	9	10	0	1	2	3	4	6	7	8	9	10	0	1	2	3	4	5

□

In Table 3.5.1, we give the parameters of some designs that can be constructed by the methods of Patterson (1952). Given p and t, these correspond to the smallest possible n for which a Patterson design may exist.

TABLE 3.5.1
Parameters of Patterson Designs

p	3	3	3	3	4	4	4	4	4
t	3	7	8	11	4	5	7	8	13
n	6	21	56	55	4	20	14	56	52
p	5	5	5	5	5	6	6	6	6
t	5	7	8	11	13	6	7	8	11
n	10	21	56	55	39	6	42	56	22

Patterson (1952) also briefly considered designs that are balanced for first and second order carryover effects and for details on these, the original source may be consulted. See also Williams (1950) in this connection.

Chapter 4

Optimal Designs via Approximate Theory

4.1 Introduction

The results discussed in Chapters 2 and 3 concern exact designs where each subject receives a sequence of treatments over p periods, and the number of subjects assigned to any such treatment sequence is a non-negative integer. In other words, with n subjects altogether, the proportion of subjects receiving any treatment sequence is of the form u/n, where $0 \leq u \leq n$. The resulting discreteness of u precludes the use of techniques based on calculus and one has to have recourse to only combinatorial arguments in the search for optimal designs. In Chapters 2 and 3, combinatorial tools were seen to yield optimal designs separately for $p \geq t$ and $p < t$. In addition, some of these results also put restrictions on the class of competing designs and most of them assume model (1.3.3) which has uncorrelated errors.

Considerable simplicity can be achieved if we allow the proportions mentioned above to vary continuously over the interval $[0, 1]$, such that their total, over all possible treatment sequences, equals unity. The resulting continuous design framework permits the development of an approximate design theory, where calculus may be used to determine these proportions in an optimal fashion. This leads to an optimal design or, more precisely, an optimal design measure. In the context of crossover designs with $t = 2$, this point was recognized, among others, by Laska, Meisner and Kushner (1983), Matthews (1987, 1990) and Kushner (1997a). A more extensive study of these designs, via the approximate theory, was reported by Kushner (1997b). He worked with arbitrary $t \geq 2$ and obtained optimality results for the estimation of direct effects under the correlated errors model (1.3.14) in the class of all connected crossover designs. Considering arbitrary t and p, which include both $p \leq t$ and $p > t$, Kushner (1997b,

1998) obtained necessary and sufficient conditions for universal optimality in terms of linear equations involving the aforesaid proportions. For $t = 2$, Matthews (1990) obtained optimal designs for direct and carry over effects using similar equations.

While the approximate theory is mathematically convenient, the ensuing optimal designs can be practically implemented for a given n only when the underlying optimal proportions are rational numbers with n as the denominator. Thus, after finding the optimal proportions for any specified t and p, it is of interest to identify the values of n for which they translate to an exact optimal design. Later in this chapter, it will be seen that symmetrized versions of approximate optimal designs play a key role in this context.

Our approach to optimality in the preceding chapters was based on the minimization of some convex and non-increasing function ϕ of the information matrix (*vide* Section 1.4). In this chapter, on the other hand, we formulate optimality in terms of maximization of concave and non-decreasing functions Φ of the information matrix. This is done primarily for keeping conformity with the literature relevant to this chapter. We, however, point out that this change is only a matter of convenience because if Φ is concave and non-decreasing then $-\Phi$ is convex and non-increasing, and thus, maximization of Φ is equivalent to minimization of $-\Phi$.

In this chapter, we follow Kushner (1997b, 1998) to give an account of the approximate theory as applied to crossover designs. In addition to the points mentioned previously, another interesting feature of his work is a general formula, pertaining to arbitrary t, n and p, for the direct effect information matrix C_{d^*} corresponding to a design d^* which is Φ-optimal in the approximate theory, where $\Phi(\cdot)$ is non-decreasing and strictly concave. For any Φ, $\Phi(C_{d^*})$ can serve as a lower bound to the Φ-value of the information matrix for direct effects of any competing design and this allows one to calculate the Φ-efficiency of direct effects for a design with the same t, n and p as d^*. Therefore, given t, n and p, if no exact optimal design can be identified either as a translation of d^* or via combinatorial arguments, then one can choose an exact design from among the available ones on the basis of their efficiencies. Moreover, if for some t, n, p, an exact design is known to be optimal in some subclass of $\Omega_{t,n,p}$, its efficiency in the global class $\Omega_{t,n,p}$ can be computed.

This chapter is organized as follows. We first follow Kushner (1997b) to present the principal results in approximate theory for the estimation of direct effects under model (1.3.14) which assumes an arbitrary dispersion matrix $\Sigma = I_n \otimes V$. Then from Section 4.4 onwards, these developments are specialized to the case $V = I_p$, as in Kushner (1998), for ease in presentation. The analogous problem of inference related to the carryover effects is briefly presented in Section 4.7. Emphasis is placed on the application of these results in the construction of exact optimal designs, both for direct and carryover effects. Some of the optimal designs discussed in the earlier chapters which arise from this approach are highlighted. Several new exact designs are also discussed. Finally, computation of the efficiency of possibly non-optimal designs is illustrated, both for direct and carryover effects.

4.2 Notation and Information Matrices

We first consider model (1.3.14) and, for introducing approximate designs in this setup, we begin with a set of treatment sequences where each sequence is a $p \times 1$ vector with elements from the set $\{1, 2, \ldots, t\}$. A typical sequence will be written as

$$\boldsymbol{s} = (t_1, t_2, \ldots, t_p)', \; t_i \in \{1, 2, \ldots, t\}, \; 1 \leq i \leq p,$$

where for $1 \leq i \leq p$, t_i denotes the treatment appearing in position i of the sequence. Let $\mathcal{S}$ be the set of all these t^p sequences. A crossover design d assigns subjects to these sequences in $\mathcal{S}$. For a given $\boldsymbol{s} \in \mathcal{S}$, let $n_{\boldsymbol{s}}$ be the number of subjects which are assigned to the sequence $\boldsymbol{s}$ and let $\sum_{\boldsymbol{s} \in \mathcal{S}}$ denote the sum over all sequences in $\mathcal{S}$. For a design d in approximate theory, the $n_{\boldsymbol{s}}$ are any numbers satisfying

$$n = \sum_{\boldsymbol{s} \in \mathcal{S}} n_{\boldsymbol{s}}, \; n_{\boldsymbol{s}} \geq 0,$$

whereas for d to be an exact design, each $n_{\boldsymbol{s}}$ needs to be an integer. Define

$$p_{\boldsymbol{s}} = n_{\boldsymbol{s}}/n, \; \boldsymbol{s} \in \mathcal{S}.$$

Thus $p_{\boldsymbol{s}}$ is the proportion of subjects assigned to $\boldsymbol{s}$ by the design d. A design in approximate theory will be specified by the set $\{p_{\boldsymbol{s}}, \boldsymbol{s} \in \mathcal{S}\}$, where

$$p_{\boldsymbol{s}} \geq 0, \; \sum_{\boldsymbol{s} \in \mathcal{S}} p_{\boldsymbol{s}} = 1. \tag{4.2.1}$$

For a sequence $\boldsymbol{s} \in \mathcal{S}$, let $T_{\boldsymbol{s}}$ be the $p \times t$ incidence matrix for the period versus direct effects and $F_{\boldsymbol{s}}$ be the $p \times t$ incidence matrix for the period versus carryover effects for all subjects assigned to $\boldsymbol{s}$. Then the information matrix $C_d(\boldsymbol{\tau}, \boldsymbol{\rho})$ under model (1.3.14) as given in (1.3.16) can be written in terms of the proportions $p_{\boldsymbol{s}}$ as shown in the following lemma.

Lemma 4.2.1.

$$C_d(\boldsymbol{\tau},\boldsymbol{\rho}) = n\sum_{\boldsymbol{s}\in\mathcal{S}} p_{\boldsymbol{s}} \begin{bmatrix} (T_{\boldsymbol{s}} - \sum_{\boldsymbol{s}\in\mathcal{S}} p_{\boldsymbol{s}}T_{\boldsymbol{s}})' \\ (F_{\boldsymbol{s}} - \sum_{\boldsymbol{s}\in\mathcal{S}} p_{\boldsymbol{s}}F_{\boldsymbol{s}})' \end{bmatrix} V^* \left[T_{\boldsymbol{s}} - \sum_{\boldsymbol{s}\in\mathcal{S}} p_{\boldsymbol{s}}T_{\boldsymbol{s}},\ F_{\boldsymbol{s}} - \sum_{\boldsymbol{s}\in\mathcal{S}} p_{\boldsymbol{s}}F_{\boldsymbol{s}} \right]$$

$$= \begin{bmatrix} C_{d11} & C_{d12} \\ C_{d21} & C_{d22} \end{bmatrix},$$

where V^* is as defined in (1.3.17).

Proof. Recalling the definitions of T_{dj} and F_{dj} from Section 1.3, from (1.3.16) we get

$$C_d(\boldsymbol{\tau},\boldsymbol{\rho}) = \begin{bmatrix} T_d' \\ F_d' \end{bmatrix} A^*[T_d, F_d] = \begin{bmatrix} \sum_{j=1}^{n} T_{dj}' A^* T_{dj} & \sum_{j=1}^{n} T_{dj}' A^* F_{dj} \\ \sum_{j=1}^{n} F_{dj}' A^* T_{dj} & \sum_{j=1}^{n} F_{dj}' A^* F_{dj} \end{bmatrix},$$

and since $A^* = H_n \otimes V^*$, the sub-matrices of $C_d(\boldsymbol{\tau},\boldsymbol{\rho})$ simplify to

$$C_{d11} = \sum_{j=1}^{n}\left(T_{dj} - n^{-1}\sum_{j=1}^{n} T_{dj}\right)' V^* \left(T_{dj} - n^{-1}\sum_{j=1}^{n} T_{dj}\right),$$

$$C_{d12} = \sum_{j=1}^{n}\left(T_{dj} - n^{-1}\sum_{j=1}^{n} T_{dj}\right)' V^* \left(F_{dj} - n^{-1}\sum_{j=1}^{n} F_{dj}\right) = C_{d21}',$$

$$C_{d22} = \sum_{j=1}^{n}\left(F_{dj} - n^{-1}\sum_{j=1}^{n} F_{dj}\right)' V^* \left(F_{dj} - n^{-1}\sum_{j=1}^{n} F_{dj}\right),$$

with V^* as in (1.3.17). It is obvious that the matrices T_{dj} and F_{dj} depend only on the sequence $\boldsymbol{s}$ to which subject j is assigned by d. So, for a design d, $T_{dj} = T_{\boldsymbol{s}}$, and $F_{dj} = F_{\boldsymbol{s}}$ for all the subjects j which are assigned to sequence $\boldsymbol{s}$. Now, since $np_{\boldsymbol{s}}$ subjects are assigned to the sequence $\boldsymbol{s}$, the lemma follows. □

The information matrices for direct and carryover effects may now be obtained from $C_d(\boldsymbol{\tau},\boldsymbol{\rho})$ as in (1.3.11).

We next introduce the notion of *symmetric* designs. To that end, we first have the following definitions. As in Section 1.4, let S_t be the symmetric group of permutations on $\{1, 2, \ldots, t\}$ and let g be any member of S_t.

Definition 4.2.1. For a sequence $\boldsymbol{s} \in \mathcal{S}$, the set of sequences $\{g\boldsymbol{s} : g \in S_t\}$ is called a symmetry block or an equivalence class and it is denoted by $< \boldsymbol{s} >$, i.e., $< \boldsymbol{s} >$ consists of all sequences obtained by relabeling of the treatments in $\boldsymbol{s}$. Sequences in the same symmetry block are called equivalent sequences.

Let L be the number of distinct treatments in $\boldsymbol{s}$. Then it is easy to see that $| < \boldsymbol{s} > | = t(t-1)\ldots(t-L+1)$, where for a set W, $|W|$ denotes its cardinality.

Example 4.2.1. With $t = 3$ and $p = 3$, consider $\boldsymbol{s} = (122)'$. Here $L = 2$ and the symmetry block $< \boldsymbol{s} >$ consists of the following six equivalent sequences: $(122)', (133)', (211)', (311)', (322)', (233)'$. □

Let $P^* = (p_{\boldsymbol{s}})$ be the array of proportions of the treatment sequences $\boldsymbol{s}$ in a design d. We can then equivalently denote the design d by the array P^* and this equivalence is denoted by $d \leftrightarrow P^*$. For a design $d \leftrightarrow P^*$, let the design obtained by relabeling the treatments of d according to g be denoted by $d_g \leftrightarrow P_g^*$. A design is called symmetric if $d = d_g$ for all $g \in S_t$. We can equivalently define a symmetric design in terms of sequence proportions as follows.

Definition 4.2.2. A symmetric design is one for which $p_{\boldsymbol{s}}$ is a constant for each $\boldsymbol{s} \in < \boldsymbol{s} >$.

In any design P^*, define

$$P_{\boldsymbol{k}} = \sum_{\boldsymbol{s} \in < \boldsymbol{k} >} p_{\boldsymbol{s}},$$

the sum being over all sequences in a symmetry block $< \boldsymbol{k} >$, $\boldsymbol{k} \in \mathcal{S}$. $P_{\boldsymbol{k}}$ is called the weight of $< \boldsymbol{k} >$ and

$$\sum_{k \in \bar{\mathcal{S}}} P_{\boldsymbol{k}} = 1 \tag{4.2.2}$$

where $\bar{\mathcal{S}}$ is the support of the design and the sum is taken over all distinct symmetry blocks in $\bar{\mathcal{S}}$. Clearly, for a symmetric design, the proportions $p_{\boldsymbol{s}}$ are given by

$$p_{\boldsymbol{s}} = \frac{P_{\boldsymbol{k}}}{| < \boldsymbol{k} > |}, \quad \boldsymbol{s} \in < \boldsymbol{k} > .$$

Example 4.2.2. With $t = 3$ and $p = 3$, consider a design supported on two symmetry blocks $< \boldsymbol{k}_1 >$ and $< \boldsymbol{k}_2 >$, with $\boldsymbol{k}_1 = (122)'$ and $\boldsymbol{k}_2 = (123)'$. Here $| < \boldsymbol{k}_1 > | = | < \boldsymbol{k}_2 > | = 6$.

If we choose $P_{\boldsymbol{k}_1} = P_{\boldsymbol{k}_2} = 1/2$, then $p_{\boldsymbol{s}} = 1/12$ for each $\boldsymbol{s} \in < \boldsymbol{k}_1 >$ and also for each $\boldsymbol{s} \in < \boldsymbol{k}_2 >$. So, with $n = 12$, the symmetric design is

$$\begin{array}{cccccccccccc} 1 & 1 & 2 & 3 & 3 & 2 & 1 & 2 & 3 & 1 & 3 & 2 \\ 2 & 3 & 1 & 1 & 2 & 3 & 2 & 1 & 2 & 3 & 1 & 3 \\ 2 & 3 & 1 & 1 & 2 & 3 & 3 & 3 & 1 & 2 & 2 & 1 \end{array}.$$

If we choose $P_{\boldsymbol{k}_1} = 1/4$ and $P_{\boldsymbol{k}_2} = 3/4$, then $p_{\boldsymbol{s}} = 1/24$ for $\boldsymbol{s} \in < \boldsymbol{k}_1 >$ and $p_{\boldsymbol{s}} = 3/24$ for $\boldsymbol{s} \in < \boldsymbol{k}_2 >$. Thus an exact symmetric design with $n = 24$ will consist of each sequence in $< \boldsymbol{k}_1 >$ repeated once and each sequence in $< \boldsymbol{k}_2 >$ repeated thrice. Again, we may choose $P_{\boldsymbol{k}_1} = 1$ and then $p_{\boldsymbol{s}} = 1/6$ for $\boldsymbol{s} \in < \boldsymbol{k}_1 >$, and $p_{\boldsymbol{s}} = 0$ otherwise. □

The above example illustrates that by varying the symmetry blocks and their weights, one can obtain different designs. For obtaining an optimal design, one has to choose an appropriate set of symmetry blocks and the corresponding weights.

We introduce some more definitions and notations which will be used in later sections.

Definition 4.2.3. A sequence $\boldsymbol{s} \in \mathcal{S}$ is called contiguous if no other treatment occurs between any two periods with the same treatment.

Definition 4.2.4. A sequence $\boldsymbol{s} \in \mathcal{S}$ is called separated if no treatment occurs in consecutive periods.

Example 4.2.3. The sequence $\boldsymbol{s} = (111122233)'$ is a contiguous sequence while $\boldsymbol{s} = (1213213)'$ is a separated sequence. □

For a given sequence $\boldsymbol{s} \in \mathcal{S}$, let

$$\begin{aligned} B_{\boldsymbol{s}} &= \text{number of consecutive treatments in } \boldsymbol{s} \\ &= |\{i \,:\, t_i = t_{i+1},\ 1 \le i \le p-1\}|, \\[1em] f_{\boldsymbol{s},\,m} &= \text{frequency of treatment } m \text{ in } \boldsymbol{s}, \\[1em] \bar{f}_{\boldsymbol{s},\,m} &= \text{frequency of treatment } m \text{ in positions} \\ &\quad\ \text{1 to } p-1 \text{ of } \boldsymbol{s}, \\[1em] f_{\boldsymbol{s},t_p} &= \text{frequency of the treatment appearing in the} \\ &\quad\ \text{last position of } \boldsymbol{s}, \\[1em] S_{\boldsymbol{s}} &= \sum_{m=1}^{t} (f_{\boldsymbol{s},\,m})^2. \end{aligned} \tag{4.2.3}$$

It is easy to see that the values of $B_{\boldsymbol{s}}, f_{\boldsymbol{s},\,t_p}$ and $S_{\boldsymbol{s}}$ remain the same for all sequences $\boldsymbol{s} \in <\boldsymbol{s}>$ and

$$B_{\boldsymbol{s}} = \begin{cases} p-L, \text{ if } \boldsymbol{s} \text{ is contiguous} \\ 0, \quad \text{if } \boldsymbol{s} \text{ is separated,} \end{cases} \tag{4.2.4}$$

$$p - L \geq B_{\boldsymbol{s}} \geq 0 \text{ for any } \boldsymbol{s} \in \mathcal{S}, \tag{4.2.5}$$

$$\sum_{m=1}^{t} f_{\boldsymbol{s},\, m} = p, \tag{4.2.6}$$

$$\sum_{m=1}^{t} \bar{f}_{\boldsymbol{s},\, m} = p-1, \tag{4.2.7}$$

where, as before, L is the number of distinct treatments in $\boldsymbol{s}$.

4.3 Quadratic Function for Direct Effects Associated with a Sequence

A quadratic function associated with a treatment sequence and the corresponding quadratic function for a design will be useful in the present context. We discuss this following Kushner (1997b).

For each treatment sequence $\boldsymbol{s} \in \mathcal{S}$, define a nonnegative quadratic $q_{\boldsymbol{s}}(u)$ as

$$q_{\boldsymbol{s}}(u) = q_{11}^{\boldsymbol{s}} + 2q_{12}^{\boldsymbol{s}}u + q_{22}^{\boldsymbol{s}}u^2, \ -\infty < u < \infty, \tag{4.3.1}$$

where

$$q_{11}^{\boldsymbol{s}} = \text{tr}[(T_{\boldsymbol{s}} - t^{-1}J_{pt})'V^*(T_{\boldsymbol{s}} - t^{-1}J_{pt})],$$

$$q_{12}^{\boldsymbol{s}} = \text{tr}[(T_{\boldsymbol{s}} - t^{-1}J_{pt})'V^*(F_{\boldsymbol{s}} - t^{-1}\bar{J}_{pt})],$$

$$q_{22}^{\boldsymbol{s}} = \text{tr}[(F_{\boldsymbol{s}} - t^{-1}\bar{J}_{pt})'V^*(F_{\boldsymbol{s}} - t^{-1}\bar{J}_{pt})],$$

$$\bar{J}_{pt} = [\mathbf{0}_t \quad J_{t,p-1}]'.$$

The quadratic $q_{\boldsymbol{s}}(u)$ as in (4.3.1) is a proper quadratic function since if $q_{22}^{\boldsymbol{s}} = 0$, then $(F_{\boldsymbol{s}} - t^{-1}\bar{J}_{pt})'V^*(F_{\boldsymbol{s}} - t^{-1}\bar{J}_{pt}) = \mathbf{0}$, which implies that

$$V^*(F_{\boldsymbol{s}} - t^{-1}\bar{J}_{pt}) = \mathbf{0} \Rightarrow (F_{\boldsymbol{s}} - t^{-1}\bar{J}_{pt}) = \mathbf{1}_p\boldsymbol{\alpha}' \text{ for some } t \times 1 \text{ vector } \boldsymbol{\alpha},$$

by (1.3.19). Since each of $F_{\boldsymbol{s}}$ and $\bar{J}_{pt}$ have $\mathbf{0}_t'$ as the first row, this implies that $\boldsymbol{\alpha} = \mathbf{0}$. Hence $F_{\boldsymbol{s}} - t^{-1}\bar{J}_{pt} = \mathbf{0}$, which is impossible.

The quadratics for sequences may in turn be used to define quadratics for designs. For a given d, $d \leftrightarrow P^*$, the quadratic of the design d, denoted by $Q(u, P^*)$, is given by

$$Q(u, P^*) = \sum_{\boldsymbol{s} \in \mathcal{S}} p_{\boldsymbol{s}} q_{\boldsymbol{s}}(u) = q_{11}(P^*) + 2q_{12}(P^*)u + q_{22}(P^*)u^2, \text{ say.} \quad (4.3.2)$$

It can be verified that the quadratics $q_{\boldsymbol{s}}(u)$ are the same for all sequences in a symmetry block $< \boldsymbol{s} >$, that is,

$$q_{\boldsymbol{s}}(u) = q_{g\boldsymbol{s}}(u), \quad g \in S_t,$$

(the verification for the case $V = I_p$ is given in Lemma 4.4.1). So instead of studying the quadratics for each individual sequence in $\mathcal{S}$, it is enough to study them for only the distinct symmetry blocks in $\mathcal{S}$.

For a symmetric design, using Lemma 4.2.1, in conjunction with (4.3.1) and (4.3.2), one can deduce the following result. As before, we write $H_t = I_t - t^{-1}J_t$.

Lemma 4.3.1. *Let $d \leftrightarrow P^*$ be a symmetric design. Then*

$$C_d(\boldsymbol{\tau}, \boldsymbol{\rho}) = \frac{n}{t-1}\begin{bmatrix} q_{11}(P^*) & q_{12}(P^*) \\ q_{12}(P^*) & q_{22}(P^*) \end{bmatrix} \otimes H_t$$

and consequently, the information matrix for direct effects is given by

$$C_d = \frac{n}{t-1}\left\{q_{11}(P^*) - q_{12}^2(P^*)/q_{22}(P^*)\right\} H_t. \quad (4.3.3)$$

□

From (4.3.2), noting that $\{q_{11}(P^*) - q_{12}^2(P^*)/q_{22}(P^*)\}$ is the minimum of $Q(u, P^*)$ over u, C_d as given in (4.3.3) may be written as

$$C_d = \frac{n}{t-1}\left\{\min_{-\infty<u<\infty} Q(u, P^*)\right\} H_t.$$

Now, let

$$b = \max_{P^*} \min_{-\infty<u<\infty} Q(u, P^*). \quad (4.3.4)$$

Kushner (1997b) showed that for any optimality functional Φ,

$$\max_d \Phi(C_d) = \frac{nb}{t-1}\Phi(H_t),$$

and that d is Φ-optimal for every strictly concave Φ and hence universally optimal, if and only if C_d satisfies

$$C_d = \frac{nb}{t-1}H_t. \quad (4.3.5)$$

It may be reiterated that this d is an optimal design in approximate theory and an exact optimal design satisfying (4.3.5) may not exist for all t, n, p. By (4.3.2) and (4.3.3), the elements of C_d involve quadratic functions of the proportions $p_{\boldsymbol{s}}$ and so it is difficult to obtain the $p_{\boldsymbol{s}}$ for the optimal design from the equations in (4.3.5). However, one can find equivalent versions of these equations which are easier to solve. In order to do this, for the set of quadratics $\{q_{\boldsymbol{s}}(u)\}$ as in (4.3.1), consider a function $q(u)$, two reals a, b and a subset $\bar{\mathcal{S}}$ of $\mathcal{S}$ as follows:

$$\begin{aligned} q(u) &= \max_{\boldsymbol{s}}\{q_{\boldsymbol{s}}(u)\}, \ -\infty < u < \infty, \\ b &= \min_{-\infty<u<\infty}\{q(u)\}, \\ a &= \arg\min_{-\infty<u<\infty}\{q(u)\}, \quad \text{i.e.,} \quad b = q(a), \\ \bar{\mathcal{S}} &= \{\boldsymbol{s} : b = q_{\boldsymbol{s}}(a), \ \boldsymbol{s} \in \mathcal{S}\}. \end{aligned} \tag{4.3.6}$$

Thus, $\bar{\mathcal{S}}$ is the set of treatment sequences in $\mathcal{S}$ at which the minimax b is achieved. Essentially using the fact that these quadratics are convex functions, it can be shown that a as in (4.3.6) is unique. Moreover, using an argument that involves interchange of maximum and minimum, Kushner (1997b) established that the b in (4.3.6) is the same as the b in (4.3.4) and that the sequences which form the support of any optimal design are contained in $\bar{\mathcal{S}}$. We skip the highly technical details and refer to Kushner (1997b) for these. Now, to obtain an optimal design one needs to determine a, b and $\bar{\mathcal{S}}$ and this is possible when V is known.

At this stage, we note that for the practically important situation of two treatments, the quadratics in (4.3.1) reduce to a simple form and consequently, it is easier to obtain optimal designs by this method. This case of $t = 2$ is discussed in Chapter 7. Furthermore, for general t, the algebra remains relatively manageable when the errors follow an autoregressive pattern and a study of this case is also deferred to Chapter 7. In this chapter, we concentrate on the case $V = I_p$ where as expected, determining a, b and $\bar{\mathcal{S}}$ for general t via (4.3.6) becomes even simpler. Thus hereafter in this chapter, we will consider the model (1.3.3). This leads to a simple set of necessary and sufficient conditions for a design to be universally optimal. These conditions are in the form of equations which are linear in the proportions and these can be solved to construct optimal designs for various choices of t and p. In Section 4.4 we present theorems under this model which will give the forms of a, b and $\bar{\mathcal{S}}$ for given values of t and p.

4.4 Determining a, b and $\bar{S}$

In the following lemma, the coefficients of the quadratic in (4.3.1) are simplified for the case $V = I_p$ and these are expressed in terms of the quantities $B_{\boldsymbol{s}}, f_{\boldsymbol{s},\, t_p}$ and $S_{\boldsymbol{s}}$ as defined in (4.2.3). Since these quantities remain constant for equivalent sequences, it follows that the quadratics are identical for all sequences belonging to the same symmetry block. This fact will be found useful later in finding optimal designs.

Lemma 4.4.1. *With $V = I_p$, the coefficients in the quadratic (4.3.1) reduce to*

$$
\begin{aligned}
q_{11}^{\boldsymbol{s}} &= p - p^{-1} S_{\boldsymbol{s}} \\
q_{12}^{\boldsymbol{s}} &= p^{-1}(pB_{\boldsymbol{s}} + f_{\boldsymbol{s},\, t_p} - S_{\boldsymbol{s}}) \qquad (4.4.1) \\
q_{22}^{\boldsymbol{s}} &= (pt)^{-1}(pt-1)(p-1) - p^{-1}(S_{\boldsymbol{s}} - 2f_{\boldsymbol{s},\, t_p} + 1).
\end{aligned}
$$

Furthermore,

$$
q'_{\boldsymbol{s}}(u) = 2p^{-1}(pB_{\boldsymbol{s}} + f_{\boldsymbol{s},\, t_p} - S_{\boldsymbol{s}}) + 2up^{-1}\{t^{-1}(pt-1)(p-1) - (S_{\boldsymbol{s}} - 2f_{\boldsymbol{s},\, t_p} + 1)\}. \qquad (4.4.2)
$$

Proof. With $V = I_p$, from (1.3.19) we have $V^* = H_p$ and hence from (4.3.1) on simplification,

$$
\begin{aligned}
q_{11}^{\boldsymbol{s}} &= \operatorname{tr}\left[T'_{\boldsymbol{s}} T_{\boldsymbol{s}} - p^{-1} T'_{\boldsymbol{s}} J_p T_{\boldsymbol{s}}\right] \\
&= \operatorname{tr}[T'_{\boldsymbol{s}} T_{\boldsymbol{s}} - p^{-1} \begin{pmatrix} f_{\boldsymbol{s},\,1} \\ \vdots \\ f_{\boldsymbol{s},\, m} \end{pmatrix} (f_{\boldsymbol{s},\,1} \cdots f_{\boldsymbol{s},\, m})] \\
&= \sum_{m=1}^{t} f_{\boldsymbol{s},\, m} - p^{-1} S_{\boldsymbol{s}} \\
&= p - p^{-1} S_{\boldsymbol{s}}.
\end{aligned}
$$

The other coefficients follow similarly. □

We now give the theorems which allow the determination of a, b and $\bar{S}$ for all t and p for the quadratics as given by Lemma 4.4.1. The proofs of the theorems use a key result from Kushner (1997a) which is stated as Lemma 4.4.2. So far, we have used the notation $\boldsymbol{s} \in \mathcal{A}$ (e.g., $\boldsymbol{s} \in \mathcal{S}$) to denote that the sequence $\boldsymbol{s}$ ranges over all sequences in a set of treatment sequences $\mathcal{A}$. In what follows, to avoid another new notation, we will also use $\boldsymbol{s} \in \mathcal{A}$ to denote that the sequence $\boldsymbol{s}$ ranges over distinct symmetry blocks in $\mathcal{A}$. The distinction between the two will be of concern only in the context of summation, where the actual meaning will be clear from the context, keeping in mind that each of $B_{\boldsymbol{s}}, S_{\boldsymbol{s}}, f_{\boldsymbol{s},t_p}$, and hence $q_{\boldsymbol{s}}(u)$, is invariant for all sequences $\boldsymbol{s}$ in a symmetry block $< \boldsymbol{s} >$.

Lemma 4.4.2. *For a fixed real number A, let*

$$\begin{aligned} B &= \max_{\boldsymbol{s}} \{q_{\boldsymbol{s}}(A)\}, \\ \mathcal{S}_A &= \{\boldsymbol{s} \,:\, q_{\boldsymbol{s}}(A) = B,\ \boldsymbol{s} \in \mathcal{S}\}, \\ \mathcal{S}_A^+ &= \{\boldsymbol{s} \,:\, q'_{\boldsymbol{s}}(A) \geq 0,\ \boldsymbol{s} \in \mathcal{S}_A\}, \ \textit{and} \\ \mathcal{S}_A^- &= \{\boldsymbol{s} \,:\, q'_{\boldsymbol{s}}(A) \leq 0,\ \boldsymbol{s} \in \mathcal{S}_A\}. \end{aligned}$$

If both sets of treatment sequences $\mathcal{S}_A^+$ and $\mathcal{S}_A^-$ are nonempty, then $a = A, b = B$ and $\bar{\mathcal{S}} = \mathcal{S}_A$. □

Theorem 4.4.1. *Suppose $p > t$. Then a, b and $\bar{\mathcal{S}}$ are given by*

$$\begin{aligned} a &= 0, \\ b &= t^{-1}p(t-1) - (pt)^{-1}r(t-r), \\ \bar{\mathcal{S}} &= \{\boldsymbol{s} \,:\, f_{\boldsymbol{s},\,m} = q \ \textit{or}\ q+1,\ 1 \leq m \leq t\}, \end{aligned} \tag{4.4.3}$$

where p is written as $p = qt + r$, and q and r are integers, $q \geq 1$, $0 \leq r \leq t-1$.

Proof. We apply Lemma 4.4.2 with $A = 0$. Towards this, we first find the possible sequences $\boldsymbol{s}$ which maximize $q_{\boldsymbol{s}}(0)$ and then consider $\mathcal{S}_0$, $\mathcal{S}_0^+$ and $\mathcal{S}_0^-$.

From Lemma 4.4.1, it can be easily seen that

$$\max_{\boldsymbol{s}} q_{\boldsymbol{s}}(0) = \max_{\boldsymbol{s}} \{p - p^{-1}S_{\boldsymbol{s}}\} = p - \min_{\boldsymbol{s}}\{p^{-1}S_{\boldsymbol{s}}\}. \tag{4.4.4}$$

It follows from Lemma 2.2.3 that the minimum of $S_{\boldsymbol{s}}$ as defined in (4.2.3), subject to (4.2.6), is attained when $\boldsymbol{s}$ is a sequence in which the $f_{\boldsymbol{s},\,m}$ values $(1 \leq m \leq t)$ differ by at most unity. Hence from (4.4.4), $q_{\boldsymbol{s}}(0)$ is maximized for sequences belonging to the set

$$\mathcal{S}_0 = \{\boldsymbol{s} \,:\, f_{\boldsymbol{s},\,m} = q \text{ or } q+1,\ 1 \leq m \leq t,\ \boldsymbol{s} \in \mathcal{S}\}.$$

Since $p = qt + r$, each sequence in $\mathcal{S}_0$ will have r treatments occurring $q+1$ times each and $t - r$ treatments occurring q times each. Therefore,

$$S_{\boldsymbol{s}} = (t-r)q^2 + r(q+1)^2 = \{p^2 + r(t-r)\}/t, \quad \boldsymbol{s} \in \mathcal{S}_0. \tag{4.4.5}$$

From (4.4.4) and (4.4.5), we also have

$$B = \max_{\boldsymbol{s}} q_{\boldsymbol{s}}(0) = p - \{p^2 + r(t-r)\}/pt \\ = t^{-1}p(t-1) - (pt)^{-1}r(t-r). \tag{4.4.6}$$

Now, with S_0 as above, we need to show that both $\mathcal{S}_0^+$ and $\mathcal{S}_0^-$ are nonempty. For this, we show that $q'_{\boldsymbol{s}}(0)$ takes both negative and nonnegative values for $\boldsymbol{s} \in \mathcal{S}_0$. From (4.4.2) and (4.2.3), for $\boldsymbol{s} \in \mathcal{S}$,

$$q'_{\boldsymbol{s}}(0) = 2p^{-1}\left(pB_{\boldsymbol{s}} + f_{\boldsymbol{s},\, t_p} - S_{\boldsymbol{s}}\right) = 2p^{-1}\left(pB_{\boldsymbol{s}} - \sum_{m=1}^{t} f_{\boldsymbol{s},\, m}\bar{f}_{\boldsymbol{s},\, m}\right). \tag{4.4.7}$$

Consider a sequence $\boldsymbol{s}_1 \in \mathcal{S}$ of the form

$$\boldsymbol{s}_1 = (1, 2, \ldots, t, 1, 2, \ldots, t, \ldots, 1, 2, \ldots, t, 1, 2, \ldots, r),' \tag{4.4.8}$$

where the string $1, 2, \ldots, t$ occurs successively q times. Clearly, $\boldsymbol{s}_1 \in \mathcal{S}_0$ and is a separated sequence. Hence from (4.2.4), $B_{\boldsymbol{s}_1} = 0$ and from (4.4.7),

$$q'_{\boldsymbol{s}_1}(0) = -2p^{-1}\sum_{m=1}^{t} f_{\boldsymbol{s}_1,\, m}\bar{f}_{\boldsymbol{s}_1,\, m} < 0.$$

Next, consider a sequence $\boldsymbol{s}_2 \in \mathcal{S}$ of the form

$$\boldsymbol{s}_2 = (1, 1, \ldots 1, 2, 2, \ldots, 2, \ldots, t, t, \ldots, t)', \tag{4.4.9}$$

where the string of identical treatments is of length q for treatments $1, 2, \ldots, (t-r)$ and is of length $q+1$ for treatments $(t-r+1), (t-r+2), \ldots, t$. Clearly, $\boldsymbol{s}_2 \in \mathcal{S}_0$ and is a contiguous sequence with $t_p = t$ and $L = t$. Hence from (4.2.4), $B_{\boldsymbol{s}_2} = p - t$ and from (4.4.5) and (4.4.7),

$$q'_{\boldsymbol{s}_2}(0) = 2p^{-1}\left\{p(p-t) + f_{\boldsymbol{s}_2,\, t} - \frac{p^2 + r(t-r)}{t}\right\}. \tag{4.4.10}$$

The values of $f_{\boldsymbol{s}_2,\, t}$ may vary depending on p, t, q and r and we consider the possible cases separately.

Case 1. $p = qt, q > 1$. Then $r = 0$ and therefore, $f_{\boldsymbol{s}_2,\, t} = q$ and (4.4.10) simplifies to $q'_{\boldsymbol{s}_2}(0) = 2t^{-1}(t-1)(p-t-1) > 0$.

Case 2. $p = t+1$. Then $q = r = 1$ and $f_{\boldsymbol{s}_2,\, t} = q + 1 = 2$. Hence from (4.4.10), $q'_{\boldsymbol{s}_2}(0) = 0$.

Case 3. $p = qt + r$, either $q = 1,\ r \geq 2$ or $q \geq 2,\ r \geq 1$. Here $f_{\boldsymbol{s}_2,\, t} = q+1$. From (4.4.10), $q'_{\boldsymbol{s}_2}(0) = 2(pt)^{-1}\{(t-1)(p-t-1)p + t - r(t-r+1)\}$. Since $p(p-t-1) > t \geq t-r+1$ and $t-1 \geq r$, it follows that $q'_{\boldsymbol{s}_2}(0) > 0$.

Thus, both $\mathcal{S}_0^-$ and $\mathcal{S}_0^+$ are nonempty and invoking Lemma 4.4.2, we get $a = A = 0$, $b = B$ and $\bar{\mathcal{S}} = \mathcal{S}_0$. From (4.4.6) and the expression for $\mathcal{S}_0$ as shown earlier, the result is now evident. □

Theorem 4.4.2. *Suppose $p \leq t$. Then a, b and $\bar{\mathcal{S}}$ are given by*

$$\begin{aligned} a &= (p-1)^{-1}, \\ b &= p - 1 - p^{-1} - \{pt(p-1)\}^{-1}, \\ \bar{\mathcal{S}} &= < \boldsymbol{s}_3 > \cup < \boldsymbol{s}_4 >, \text{ where} \\ \boldsymbol{s}_3 &= (1, 2, \ldots p-1, p)', \quad \boldsymbol{s}_4 = (1, 2, \ldots p-1, p-1)'. \end{aligned} \tag{4.4.11}$$

Proof. The proof follows along the line of that of Theorem 4.4.1. However, now we apply Lemma 4.4.2 with $A = (p-1)^{-1}$. Lemma 4.4.1 yields

$$q_{\boldsymbol{s}}(A) = \frac{2}{p-1} B_{\boldsymbol{s}} - \frac{p}{(p-1)^2} S_{\boldsymbol{s}} + \frac{2}{(p-1)^2} f_{\boldsymbol{s},\, t_p} + b_0 + \frac{p^2-2}{(p-1)^2}, \tag{4.4.12}$$

for all $\boldsymbol{s} \in \mathcal{S}$, and where $b_0 = p - 1 - p^{-1} - \{pt(p-1)\}^{-1}$.

Given a sequence $\boldsymbol{s}$, suppose we derive another sequence $\boldsymbol{k}$ from it by keeping the treatment in position p of $\boldsymbol{s}$ fixed and rearranging the treatments in the other positions. Then $S_{\boldsymbol{k}} = S_{\boldsymbol{s}}$ and $f_{\boldsymbol{k},\, t_p} = f_{\boldsymbol{s},\, t_p}$ for all such $\boldsymbol{k}$ though $B_{\boldsymbol{k}}$ may not equal $B_{\boldsymbol{s}}$. So, noting that $B_{\boldsymbol{s}}$ appears in (4.4.12) with a positive coefficient and recalling the facts (4.2.4) and (4.2.5) about $B_{\boldsymbol{s}}$, it is evident that a sequence maximizing $q_{\boldsymbol{s}}(A)$ must be a contiguous one. Continuing with $A = (p-1)^{-1}$, for any contiguous sequence $\boldsymbol{s}$ with L distinct treatments, from (4.2.4) and (4.4.12), we observe that

$$(p-1)^2 q_{\boldsymbol{s}}(A) = -pS_{\boldsymbol{s}} + 2f_{\boldsymbol{s},\, t_p} - 2(p-1)L + b_0(p-1)^2 + 3p^2 - 2p - 2. \tag{4.4.13}$$

Now, given a contiguous sequence $\boldsymbol{s}$, if we relabel the treatments in $\boldsymbol{s}$ to obtain a new sequence $\boldsymbol{k}_1$ then, $S_{\boldsymbol{k}_1} = S_{\boldsymbol{s}}$, while $f_{\boldsymbol{k}_1,\, t_p} = f_{\boldsymbol{s},\, m}$ for some m such that $f_{\boldsymbol{s},\, m} \geq 1$. Hence from (4.4.13), a contiguous sequence $\boldsymbol{s}$ maximizing $q_{\boldsymbol{s}}(A)$ must satisfy

$$f_{\boldsymbol{s},\, t_p} = \max\{f_{\boldsymbol{s},\, m}\ ,\ 1 \leq m \leq t\}. \tag{4.4.14}$$

Clearly, with $\boldsymbol{s}_3$ and $\boldsymbol{s}_4$ defined as in (4.4.11), $< \boldsymbol{s}_3 >$ consists of all sequences having distinct treatments in the p positions and $< \boldsymbol{s}_4 >$ consists of all sequences having distinct treatments in the first $p-1$ positions with the treatment in position $p-1$ also occurring in position p. Kushner (1998) argued that for the maximization of $q_{\boldsymbol{s}}(A)$, one must have

$$\begin{aligned} &\text{either (i) } f_{\boldsymbol{s},\, t_p} = 1 \text{ and hence } \boldsymbol{s} \in < \boldsymbol{s}_3 >, \\ &\text{or} \quad \text{(ii) } f_{\boldsymbol{s},\, t_p} = 2 \text{ and } \boldsymbol{s} \in < \boldsymbol{s}_4 > . \end{aligned}$$

If (i) holds then $L = p$, $S_{\boldsymbol{s}} = p$ and hence by (4.4.13), $q_{\boldsymbol{s}}(A) = b_0$, on simplification. Similarly, if (ii) holds then $L = p - 1$, $S_{\boldsymbol{s}} = p + 2$ and again (4.4.13) yields $q_{\boldsymbol{s}}(A) = b_0$. Thus $q_{\boldsymbol{s}}(A)$ is maximized if and only if $\boldsymbol{s} \in < \boldsymbol{s}_3 >$ or $\boldsymbol{s} \in < \boldsymbol{s}_4 >$, i.e.,

$$\mathcal{S}_A = < \boldsymbol{s}_3 > \cup < \boldsymbol{s}_4 >,$$

and the maximum value of $q_{\boldsymbol{s}}(A)$, that is B, equals b_0.

Finally, with $\mathcal{S}_A$ as above, we need to show that both $\mathcal{S}_A^+$ and $\mathcal{S}_A^-$ are nonempty. Using (4.4.2), one can see that,

$$\begin{aligned} q'_{\boldsymbol{s}}(A) &= -2(pt)^{-1} < 0 \text{ for } \boldsymbol{s} \in < \boldsymbol{s}_3 > \\ &= 2(pt)^{-1}\{t(p-1)-1\} > 0 \text{ for } \boldsymbol{s} \in < \boldsymbol{s}_4 > . \end{aligned} \qquad (4.4.15)$$

Thus both $\mathcal{S}_A^+$ and $\mathcal{S}_A^-$ are nonempty. Invoking Lemma 4.4.2, we get $a = A = (p-1)^{-1}$, $b = B = b_0$ and $\bar{\mathcal{S}} = \mathcal{S}_A$. From the expressions for b_0 and $\bar{\mathcal{S}}_A$ shown above, it is clear that a, b and $\bar{\mathcal{S}}$ are as specified by (4.4.11).

□

Using a, b and $\bar{\mathcal{S}}$ from the above two theorems, one can find optimal designs as shown in the next two sections.

4.5 Optimality Equations

From (4.3.5) and Theorems 4.4.1 and 4.4.2 the next theorem follows.

Theorem 4.5.1. *The information matrix for direct effects for a universally optimal design d is given by*

$$C_d = \frac{nb}{t-1} H_t \qquad (4.5.1)$$

where b is as in (4.4.3) if $p > t$ and as in (4.4.11) if $p \leq t$. □

It is difficult to solve the equations in (4.5.1) directly to obtain the proportions $p_{\boldsymbol{s}}$ for the optimal design, but these may be reduced to a simpler system of equations which give the necessary and sufficient conditions for optimality. In order to state these optimality equations we need some notations which we introduce below following Kushner (1998). Let

$$B_{lm} = \sum_{i=1}^{p-1} \sum_{\boldsymbol{s} \in \mathcal{S}: t_{i+1}=l, t_i=m} p_{\boldsymbol{s}}, \; 1 \leq m, l \leq t,$$

$$P_m = \sum_{\boldsymbol{s} \in \mathcal{S}} p_{\boldsymbol{s}} f_{\boldsymbol{s}, m}, \; 1 \leq m \leq t,$$

$$R_m = \sum_{s \in \mathcal{S}} p_s \bar{f}_{s,\,m},\ 1 \le m \le t,$$

$$P_{ml} = \sum_{s \in \mathcal{S}} p_s f_{s,\,m} f_{s,\,l},\ 1 \le m, l \le t,$$

$$Q_{ml} = \sum_{s \in \mathcal{S}} p_s f_{s,\,m} \bar{f}_{s,\,l},\ 1 \le m, l \le t,$$

$$R_{ml} = \sum_{s \in \mathcal{S}} p_s \bar{f}_{s,\,m} \bar{f}_{s,\,l},\ 1 \le m, l \le t,$$

$$P_m^i = \sum_{s \in \mathcal{S}: t_i = m} p_s,$$

$$\delta_m^l = \begin{cases} 1, & \text{if } m = l \\ 0, & \text{otherwise.} \end{cases}$$

It may be pointed out that for an exact design d, $nB_{lm} = z_{dlm}$, $nP_m = r_{dm}$, $nR_m = \bar{r}_{dm}$, and so on, where $z_{dlm}, r_{dm}, \bar{r}_{dm}$, etc., are as defined in (1.3.1).

Theorem 4.5.2. *When $p > t$, and b is as in (4.4.3), a design d is universally optimal for direct effects if and only if its support is contained in $\bar{S}$ as in (4.4.3) and for $1 \le m, l \le t$, $1 \le i \le p$,*

$$(i)\quad \delta_m^l P_m - (1/p) P_{ml} = b(t\delta_m^l - 1)/(t^2 - t),$$

$$(ii)\quad B_{lm} = (1/p) Q_{lm},$$

$$(iii)\quad P_m^i = 1/t.$$

□

Theorem 4.5.3. *When $2 < p \le t$, and b is as in (4.4.11), a design d is universally optimal for direct effects if and only if its support is contained in $\bar{S}$ as given in (4.4.11) and for $1 \le m, l \le t$, $1 \le i \le p$,*

$$(i)\quad \delta_m^l P_m - p^{-1} P_{ml} + (p-1)^{-1}(B_{ml} - p^{-1} Q_{ml}) = b(t\delta_m^l - 1)/(t^2 - t),$$

$$(ii)\quad B_{lm} - p^{-1} Q_{lm} + (p-1)^{-1}(\delta_m^l R_m - p^{-1} R_{ml}) = 1/(pt^2),$$

$$(iii)\quad P_m^i = 1/t.$$

□

The above theorems pertain to approximate design theory and the conditions given there may not be satisfied by a universally optimal design in exact design theory. The optimality equations of Theorems 4.5.2 and 4.5.3 can be solved by numerical methods and Kushner (1998) solved them to obtain universally optimal designs in approximate theory for the cases $(t,p) = (3,3), (4,3), (4,4)$ and $(3,5)$. The proportions used in each of these designs are rationals and these designs are also optimal in exact theory with the minimum numbers of subjects required for the corresponding exact designs being $12, 32, 144$ and 30, respectively. The optimal design for $(t,p) = (3,3)$ and $n = 12$ is the design in Example 4.2.2.

Remark 4.5.1. Condition (iii) in each of the above two theorems is equivalent to the condition of uniformity on periods, i.e., such uniformity is necessary for universal optimality in approximate theory. □

Remark 4.5.2. If $p \geq t$, and $t|p$, then (4.4.3) implies that $f_{\boldsymbol{s},m} = p/t$ for all treatments m. This is the same as saying that the design is uniform on subjects. So, for $p \geq t$, and $t|p$, uniformity on subjects is also a necessary condition for universal optimality in approximate theory. □

Remark 4.5.3. Even in exact theory, the types of uniformity noted in the last two remarks play a crucial rôle in ensuring universal optimality. This is evident, for instance, from Theorems 2.3.3, 2.3.4, 2.4.1, 2.5.4, 3.2.4, etc. □

Remark 4.5.4. It can be shown that some of the universally optimal exact designs of earlier chapters, e.g., the optimal designs in Theorem 2.4.1, Theorem 2.4.2 and Theorem 3.2.4 (with $(p-1)t|n$)), satisfy the conditions of the above theorems. Hence they are universally optimal for direct effects in approximate design theory. □

4.6 Optimal Symmetric Designs for Direct Effects

Using concavity arguments, it can be seen that among the optimal designs there is always one that is symmetric (*cf.* Matthews (1987), Kushner (1997b)). If we restrict to optimal symmetric designs, then the conditions for universal optimality become much simpler than those in the theorems of the preceding section. The resulting reduced conditions, appearing in the following theorems, are useful in constructing optimal symmetric designs.

Theorem 4.6.1. *(i) A design is universally optimal for direct effects only if its support is in $\bar{S}$ and the condition*

$$\sum_{\boldsymbol{s}\in\bar{S}} P_{\boldsymbol{s}} q'_{\boldsymbol{s}}(a) = 0 \tag{4.6.1}$$

holds, where $P_{\boldsymbol{s}}$ is the weight of the symmetry block $< \boldsymbol{s} >$, a, $\bar{S}$ are as in (4.3.6), $q'_{\boldsymbol{s}}(a)$ is given by (4.4.2) and the summation is over all distinct symmetry blocks in $\bar{S}$.
(ii) Furthermore, a symmetric design is universally optimal for direct effects if and only if its support is in $\bar{S}$ and (4.6.1) holds. □

Two illustrative examples follow.

Example 4.6.1. Let $p = 4$, $t = 2$. From Theorem 4.4.1, $a = 0$ and $\bar{S}$ consists of symmetry blocks of sequences $(1122)', (1212)'$ and $(1221)'$. On computation using (4.4.7), the $q'_{\boldsymbol{s}}(0)$ values for these sequences are found to be $1, -3$ and -1, respectively. So, a universally optimal design may be constructed by choosing suitable symmetry blocks and their weights such that (4.6.1) is satisfied. The following are some of the possible choices. These optimal designs were also obtained by Matthews (1990).

(i)$P_{(1122)'} = P_{(1221)'} = 1/2, P_{(1212)'} = 0$. This leads to $p_{\boldsymbol{s}} = 1/4$ for $\boldsymbol{s} \in< (1122)' > \cup < (1221)' >$ and all other $p_{\boldsymbol{s}} = 0$. The corresponding smallest exact design has $n = 4$.

(ii) $P_{(1122)'} = 4/6, P_{(1212)'} = P_{(1221)'} = 1/6$. This leads to $p_{\boldsymbol{s}} = 4/12$ for $\boldsymbol{s} \in< (1122)' >$ and $p_{\boldsymbol{s}} = 1/12$ for $\boldsymbol{s} \in< (1212)' > \cup < (1221)' >$. The corresponding smallest exact design has $n = 12$.

(iii) $P_{(1122)'} = 3/4, P_{(1212)'} = 1/4, P_{(1221)'} = 0$. This leads to $p_{\boldsymbol{s}} = 3/8$ for $\boldsymbol{s} \in< (1122)' >$, $p_{\boldsymbol{s}} = 1/8$ for $\boldsymbol{s} \in< (1212)' >$ and all other $p_{\boldsymbol{s}} = 0$. The corresponding smallest exact design has $n = 8$. □

Example 4.6.2. Let $p = 3$, $t = 4$. From Theorem 4.4.2, $a = 1/2$ and $\bar{S} =< (123)' > \cup < (122)' >$. From (4.4.14), $q'_{\boldsymbol{s}}(1/2) = -1/6$ for $\boldsymbol{s} \in< (123)' >$ while $q'_{\boldsymbol{s}}(1/2) = 7/6$ for $\boldsymbol{s} \in< (122)' >$. Therefore, by Theorem 4.6.1, a universally optimal design may be constructed by choosing $P_{(123)'} = 7/8$ and $P_{(122)'} = 1/8$, leading to $p_{\boldsymbol{s}} = 7/48$ for $\boldsymbol{s} \in< (123)' >$ and $p_{\boldsymbol{s}} = 1/48$ for $\boldsymbol{s} \in< (122)' >$. The corresponding smallest exact design has $n = 48$. □

This idea of construction in Examples 4.6.1 and 4.6.2 may be used to derive simpler necessary and sufficient conditions for a symmetric design to be universally optimal. These conditions arising from (4.6.1) do not involve explicit evaluation of $q'_{\boldsymbol{s}}(a)$ and are summarized in the next two theorems.

Theorem 4.6.2. *When $p > t$, a symmetric design is universally optimal for direct effects if and only if*

$$(i)\ \sum_{\boldsymbol{s}\in\bar{\mathcal{S}}_q} (pB_{\boldsymbol{s}} - w)\,P_{\boldsymbol{s}} + \sum_{\boldsymbol{s}\in\bar{\mathcal{S}}_{q+1}} (pB_{\boldsymbol{s}} - w + 1)\,P_{\boldsymbol{s}} = 0 \ \ and$$

$$(ii)\ \sum_{\boldsymbol{s}\in\bar{\mathcal{S}}} P_{\boldsymbol{s}} = 1,$$

where $\bar{\mathcal{S}}, q$ and r are as in Theorem 4.4.1, $w = \{p(p-1) + r(t-r+1)\}/t$ and we write $\bar{\mathcal{S}} = \bar{\mathcal{S}}_q \cup \bar{\mathcal{S}}_{q+1}$ with $\bar{\mathcal{S}}_q = \{\boldsymbol{s} : \boldsymbol{s} \in \bar{\mathcal{S}},\ f_{\boldsymbol{s},t_p} = q\}$, $\bar{\mathcal{S}}_{q+1} = \{\boldsymbol{s} : \boldsymbol{s} \in \bar{\mathcal{S}},\ f_{\boldsymbol{s},t_p} = q+1\}$.

Proof. On simplification from (4.4.7) using (4.4.5),

$$q'_{\boldsymbol{s}}(0) = (2/p)(pB_{\boldsymbol{s}} - w), \text{ for } \boldsymbol{s} \in \bar{\mathcal{S}}_q$$

and

$$q'_{\boldsymbol{s}}(0) = (2/p)(pB_{\boldsymbol{s}} - w + 1) \text{ for } \boldsymbol{s} \in \bar{\mathcal{S}}_{q+1}.$$

Now the theorem follows from (4.6.1). □

Theorem 4.6.3. *When $p \le t$, a symmetric design is universally optimal for direct effects if and only if*

$$P_{\boldsymbol{s}_3} = 1 - \frac{1}{t(p-1)} \text{ and } P_{\boldsymbol{s}_4} = \frac{1}{t(p-1)},$$

where sequences $\boldsymbol{s}_3, \boldsymbol{s}_4$ are as in (4.4.11).

Proof. From (4.4.15) and (4.6.1),

$$-P_{\boldsymbol{s}_3} + \{t(p-1) - 1\}P_{\boldsymbol{s}_4} = 0.$$

Since $P_{\boldsymbol{s}_3} + P_{\boldsymbol{s}_4} = 1$, the result follows. □

A version of Theorem 4.6.2 for $t = 2$ was given by Matthews (1990). By choosing symmetry blocks and their weights appropriately such that the conditions of the above two theorems are satisfied, various symmetric universally optimal designs may be constructed. These may in turn be used to construct symmetric universally optimal designs with suitable n in exact design theory. Thus these exact optimal designs exist for all t and p. The only drawback of this method is that sometimes n can become large. Moreover, as the size of a symmetry block $< \boldsymbol{s} >$ depends on the number of distinct treatments in $\boldsymbol{s}$, the value of n depends on the symmetry blocks chosen as the support of the design. Therefore, a judicious choice of these symmetry blocks is important.

We now present some illustrative examples. In these examples we only give the nonzero values of $P_{\boldsymbol{s}}$ and $p_{\boldsymbol{s}}$. For all other symmetry blocks in $\bar{\mathcal{S}}$, $P_{\boldsymbol{s}} = 0$ and consequently, $p_{\boldsymbol{s}} = 0$ for all sequences in these symmetry blocks.

Example 4.6.3. Let $p = t = 3$. Consider the sequences $\boldsymbol{s}_3 = (123)'$ and $\boldsymbol{s}_4 = (122)'$. By Theorem 4.6.3 a universally optimal design has $P_{\boldsymbol{s}_3} = 5/6$ and $P_{\boldsymbol{s}_4} = 1/6$. Since both the symmetry blocks consist of 6 equivalent sequences, this design has $p_{\boldsymbol{s}} = 5/36$ for $\boldsymbol{s} \in < \boldsymbol{s}_3 >$ and $p_{\boldsymbol{s}} = 1/36$ for $\boldsymbol{s} \in < \boldsymbol{s}_4 >$. The corresponding smallest exact design has $n = 36$. □

Example 4.6.4. Let $p = 6, t = 3$. We apply Theorem 4.6.2. Here $q = 2, r = 0$ and so, $w = 10$. For illustration, we consider the following sequences:

$$\boldsymbol{k}_1 = (112233)', \ \boldsymbol{k}_2 = (122331)', \ \boldsymbol{k}_3 = (121332)', \ \boldsymbol{k}_4 = (123123)'.$$

All these sequences are in $\bar{\mathcal{S}}_q$. Taking zero weights for all symmetry blocks in $\bar{\mathcal{S}}$ other than those for $\boldsymbol{k}_1, \boldsymbol{k}_2, \boldsymbol{k}_3, \boldsymbol{k}_4$, condition (i) of Theorem 4.6.2 reduces to

$$8P_{\boldsymbol{k}_1} + 2P_{\boldsymbol{k}_2} - 4P_{\boldsymbol{k}_3} - 10P_{\boldsymbol{k}_4} = 0.$$

Different solutions of this equation which satisfy condition (ii) of Theorem 4.6.2 will give different symmetric universally optimal designs. A few of them are shown below.

1. Take $P_{\boldsymbol{k}_1} = 1/3, P_{\boldsymbol{k}_3} = 2/3$. Then the proportions for the universal optimal design are $p_{\boldsymbol{s}} = 1/18$ for $\boldsymbol{s} \in < \boldsymbol{k}_1 >$ and $p_{\boldsymbol{s}} = 1/9$ for $\boldsymbol{s} \in < \boldsymbol{k}_3 >$. The corresponding smallest exact design has $n = 18$.

2. Take $P_{\boldsymbol{k}_1} = P_{\boldsymbol{k}_2} = P_{\boldsymbol{k}_4} = 1/3$. Then the proportions for the universal optimal design are $p_{\boldsymbol{s}} = 1/18$ for $\boldsymbol{s} \in < \boldsymbol{k}_1 > \cup < \boldsymbol{k}_2 > \cup < \boldsymbol{k}_4 >$. The corresponding smallest exact design has $n = 18$.

3. Take $P_{\boldsymbol{k}_2} = 2/3$ and $P_{\boldsymbol{k}_3} = 1/3$. Then the proportions for the optimal design are $p_{\boldsymbol{s}} = 1/9$ for $\boldsymbol{s} \in < \boldsymbol{k}_2 >$ and $p_{\boldsymbol{s}} = 1/18$ for $\boldsymbol{s} \in < \boldsymbol{k}_3 >$. The corresponding smallest exact design has $n = 18$.

4. Take $P_{\boldsymbol{k}_2} = 5/6$ and $P_{\boldsymbol{k}_4} = 1/6$. Then the proportions for the optimal design are $p_{\boldsymbol{s}} = 5/36$ for $\boldsymbol{s} \in < \boldsymbol{k}_2 >$ and $p_{\boldsymbol{s}} = 1/36$ for $\boldsymbol{s} \in < \boldsymbol{k}_4 >$. The corresponding smallest exact design has $n = 36$. □

Example 4.6.5. Let $p = 4, t = 3$. We again apply Theorem 4.6.2. Here $q = 1, r = 1$ and so, $w = 5$. From Theorem 4.6.2, the coefficient of $P_{(1233)'}$ in (i) is zero. Therefore, the design with $P_{(1233)'} = 1$ will be universally optimal and it will have proportions $p_{\boldsymbol{s}} = 1/6$ for $\boldsymbol{s} \in <(1233)'>$. The corresponding smallest exact design has $n = 6$. In fact, for all p, t with $p = t + 1$, the coefficient of $P_{\boldsymbol{s}_2}$ in condition (i) of Theorem 4.6.2 is zero, where

$$\boldsymbol{s}_2 = (1, 2, \ldots, t-1, t, t)'.$$

So a design with $P_{\boldsymbol{s}_2} = 1$ will be universally optimal for all such pairs p, t.

□

Remark 4.6.1. (i) The above examples demonstrate the flexibility of the present method. However, for larger values of t, the number of subjects required in the corresponding exact design can be quite large. The first three designs of Example 4.6.4 all have $n = 18$ and the fourth design requires more subjects due to the choice of the symmetry blocks used in its construction. Thus one has to choose the symmetry blocks judiciously while constructing an optimal design so as to keep the number of subjects under control as much as possible. In this connection, we refer to Kushner (1999) who proposed a method that, in a sense, exploits partitioned versions of the symmetry blocks so as to achieve optimality, keeping the size of the experiment relatively small.

(ii) The optimal designs for $t = 2$ which may be obtained by the above method are the same as those obtained by Matthews (1990). Designs in Example 4.6.5 are the extra-period crossover designs which were shown to be universally optimal by Cheng and Wu (1980) and are discussed in Theorem 2.4.2 in Chapter 2. □

4.7 Optimal Designs for Carryover Effects

Analogous to the discussion in the earlier sections where the focus was on inference on direct effects, results may be derived for inference on carryover effects and universally optimal designs for arbitrary p and t can be obtained, without placing restrictions on the class of competing designs.

To begin with, for a sequence $\boldsymbol{s} \in \bar{S}$, we define the following quadratics for carryover effects:

$$r_{\boldsymbol{s}}(u) = q_{22}^{\boldsymbol{s}} + 2q_{12}^{\boldsymbol{s}}u + q_{11}^{\boldsymbol{s}}u^2, \ -\infty < u < \infty, \tag{4.7.1}$$

where the coefficients are as defined in (4.4.1). Note that the roles of the coefficient of u^2 and the constant term in (4.7.1) and (4.3.1) have been interchanged. Consequently, in order that (4.7.1) is a proper quadratic, we now must have $q_{11}^{\boldsymbol{s}} \neq 0$. From the expression of $q_{11}^{\boldsymbol{s}}$, it can easily be argued that $q_{11}^{\boldsymbol{s}} = 0$ if and only if the same treatment occurs in every position of $\boldsymbol{s}$. Henceforth, such sequences are kept out of consideration and in view of this, we consider the class of sequences

$$\mathcal{S}^* = \{\boldsymbol{s} = (t_1, \ldots, t_p)', \ \ t_1, \ldots, t_p \text{ are not all equal}\}.$$

Let

$$\begin{aligned} r(u) &= \max_{\boldsymbol{s}}\{r_{\boldsymbol{s}}(u)\}, \ -\infty < u < \infty, \\ \bar{b} &= \min_{-\infty<u<\infty} \{r(u)\}, \\ \bar{a} &= \arg \min_{-\infty<u<\infty} \{r(u)\}, \ \ i.e., \ \ \bar{b} = r(\bar{a}), \\ \bar{\mathcal{S}}^* &= \{\boldsymbol{s} : \bar{b} = r_{\boldsymbol{s}}(\bar{a}), \ \boldsymbol{s} \in \mathcal{S}^*\}. \end{aligned} \tag{4.7.2}$$

We can determine $\bar{a}, \bar{b}$ and $\bar{\mathcal{S}}^*$ applying a counterpart of Lemma 4.4.2 where $q_{\boldsymbol{s}}(\cdot)$ is replaced by $r_{\boldsymbol{s}}(\cdot)$, and results for inference on carryover effects, analogous to the theorems in Sections 4.4–4.6, can be derived along similar lines. We skip the details, giving only the result corresponding to Theorem 4.6.1 and illustrating its use in the construction of optimal designs through some examples. For details we refer to Bose and Shah (2005).

Theorem 4.7.1. *(i) A design is universally optimal for carryover effects only if its support is in $\bar{\mathcal{S}}^*$ and the condition*

$$\sum_{\boldsymbol{s} \in \bar{\mathcal{S}}^*} P_{\boldsymbol{s}} r'_{\boldsymbol{s}}(\bar{a}) = 0 \tag{4.7.3}$$

holds, where $P_{\boldsymbol{s}}$ is the weight of the symmetry block $< \boldsymbol{s} >$, $\bar{a}$, $\bar{\mathcal{S}}^*$ are as in (4.7.2), $r'_{\boldsymbol{s}}(\bar{a})$ is as obtained from (4.7.1) and the summation is over all distinct symmetry blocks in $\bar{\mathcal{S}}^*$.
(ii) Furthermore, a symmetric design is universally optimal for carryover effects if and only if its support is in $\bar{\mathcal{S}}^*$ and (4.7.3) holds. □

It can be shown that the version of Lemma 4.4.2, with $r_{\boldsymbol{s}}(\cdot)$ replacing $q_{\boldsymbol{s}}(\cdot)$, is applicable to every p and t with $A = 0$, resulting in $\bar{a} = 0$ (this is in contrast with the situation for direct effects where the cases $p > t$ and

$p \leq t$ had to be studied separately). Thus for every p and t, from (4.7.1) and (4.4.1), we have

$$r_{\boldsymbol{s}}(0) = (pt)^{-1}(pt-1)(p-1) - p^{-1}(S_{\boldsymbol{s}} - 2f_{\boldsymbol{s},\, t_p} + 1) \tag{4.7.4}$$

and

$$r'_{\boldsymbol{s}}(0) = 2p^{-1}(pB_{\boldsymbol{s}} + f_{\boldsymbol{s},\, t_p} - S_{\boldsymbol{s}}). \tag{4.7.5}$$

From (4.7.4), it is clear that in order to maximize $r_{\boldsymbol{s}}(0)$, one needs to minimize $S_{\boldsymbol{s}} - 2f_{\boldsymbol{s},\, t_p}$. The following exhaustive cases summarize the final results so obtained.

Case 1: $p \leq t$. Here

$$\bar{b} = (pt)^{-1}(pt-t-1)(p-1), \text{ and}$$
$$\bar{\mathcal{S}}^* = <\boldsymbol{s}_3> \cup <\boldsymbol{s}_4> \cup \Delta_1$$

where sequences $\boldsymbol{s}_3$ and $\boldsymbol{s}_4$ are as in (4.4.11) and Δ_1 is the union of distinct symmetry blocks arising out of the sequences in

$$\mathcal{T}_1 = \{\boldsymbol{s} : \boldsymbol{s} = (t_1, \ldots, t_p)',\ t_i = t_p \text{ for some } i < p-1 \text{ and } t_1, \ldots, t_{p-1} \text{ all distinct}\}.$$

Example 4.7.1. Let $p = t = 3$. Here, $\boldsymbol{s}_3 = (123)', \boldsymbol{s}_4 = (122)'$ and by (4.7.5), $r'_{\boldsymbol{s}}(0)$ equals $-4/3$ and 0 for $\boldsymbol{s} = \boldsymbol{s}_3$ and $\boldsymbol{s}_4$, respectively. Moreover, $r'_{\boldsymbol{s}}(0) = -2$ for every $\boldsymbol{s} \in \mathcal{T}_1$. Hence from (4.7.3), the optimal design assigns the entire weight on $<(122)'>$. It is not hard to see that the same happens when $p = 3$ and $t = 4$. □

Case 2: $p = qt, q > 1$. Here

$$\bar{b} = (pt)^{-1}(p^2 - p - 1)(t-1) \text{ and}$$
$$\bar{\mathcal{S}}^* = \Delta_2 \cup \Delta_3,$$

where Δ_u is the union of distinct symmetry blocks arising out of the sequences in $\mathcal{T}_u$, $u = 2, 3$, and

$$\mathcal{T}_2 = \{\boldsymbol{s} : \text{ each treatment occurs } q \text{ times in } \boldsymbol{s}\},$$
$$\mathcal{T}_3 = \{\boldsymbol{s} : f_{\boldsymbol{s},\, t_p} = q+1, \text{one treatment other than } t_p \text{ occurs } q-1 \text{ times and the remaining } t-2 \text{ treatments occur } q \text{ times each in } \boldsymbol{s}\}.$$

Example 4.7.2. Let $p = 4, t = 2$. Explicit consideration of Δ_2 and Δ_3 show that here $\bar{\mathcal{S}}^*$ consists of symmetry blocks of sequences

$(1122)', (1212)', (1221)', (1222)', (1211)', (1121)'$ with their $r'_{\boldsymbol{s}}(0)$ values being $1, -3, -1, 1/2, -3/2, -3/2$, respectively. Equation (4.7.3) is met, for instance, when (i) $P_{(1122)'} = P_{(1221)'} = 1/2$, or (ii) $P_{(1122)'} = 3/4, P_{(1212)'} = 1/4$. These designs were obtained by Matthews (1990). □

Example 4.7.3. Let $p = 6, t = 3$. As before, one choice of an optimal design consists of symmetry blocks $< (112233)' >$, $< (123123)' >$, $< (122331)' >$ applied in equal proportions. This follows on noting that the $r'_{\boldsymbol{s}}(0)$ values for sequences in these symmetry blocks equal $8/3, -10/3$ and $2/3$, respectively. □

Case 3: $p = qt + 1$. Here

$$\bar{b} = t^{-1}(p-1)(t-1) \text{ and}$$
$$\bar{\mathcal{S}}^* = \Delta_4,$$

where Δ_4 is the union of distinct symmetry blocks arising out of the sequences in

$$\mathcal{T}_4 = \{\boldsymbol{s} : f_{\boldsymbol{s},t_p} = q+1, \text{ the remaining } t-1 \text{ treatments occur } q \text{ times each in } \boldsymbol{s}\}.$$

Example 4.7.4. Let $p = 4, t = 3$. Note that the sequence $(1233)'$ belongs to $\bar{\mathcal{S}}^*$ and that its $r'_{\boldsymbol{s}}(0)$ value equals 0. So the optimal design assigns the entire weight on the symmetry block $< (1233)' >$. □

Case 4: $p = qt + r$, $2 \leq r < t$. Here

$$\bar{b} = (pt)^{-1}\{(p-1)(pt-1) - (p-r)(p+r-2) - t(r-1)\} \text{ and}$$
$$\bar{\mathcal{S}}^* = \Delta_5 \cup \Delta_6,$$

where Δ_u is the union of distinct symmetry blocks arising out of the sequences in $\mathcal{T}_u$, $u = 5, 6$ and

$$\mathcal{T}_5 = \{\boldsymbol{s} : f_{\boldsymbol{s},\, t_p} = q+1,\ r-1 \text{ treatments other than } t_p \text{ also occur } q+1 \text{ times each and the remaining } t-r \text{ treatments occur } q \text{ times each in } \boldsymbol{s}\},$$
$$\mathcal{T}_6 = \{\boldsymbol{s} : f_{\boldsymbol{s},\, t_p} = q+2,\ (r-2) \text{ treatments occur } q+1 \text{ times each and the remaining } t-r+1 \text{ treatments occur } q \text{ times each in } \boldsymbol{s}\}.$$

Example 4.7.5. Let $p = 5, t = 3$. Here one choice of the optimal design consists of the symmetry blocks $< (11232)' >$ and $< (12333)' >$ applied in equal proportions. □

Remark 4.7.1. As noted in Example 4.6.1, the two designs in Example 4.7.2 are also universally optimal for direct effects. Design (i) of Example 4.7.2 is a strongly balanced uniform design. Such designs are known to be universally optimal for both direct and carryover effects (*cf.* Theorem 2.4.1). Again, as noted in Example 4.6.5, the design in Example 4.7.4 is also universally optimal for direct effects. This is an extra-period strongly balanced design, which is known to be optimal for both direct and carryover effects (*cf.* Theorem 2.4.2). The designs in Example 4.7.1 may also be constructed by the method proposed by Stufken (1991), who showed that these designs are universally optimal for carryover effects. □

4.8 Design Efficiency

As seen in the preceding sections, in many situations, universally optimal designs obtained via approximate theory lend themselves to universally optimal exact designs for appropriately chosen n. Even when for a pre-specified n, this kind of implementation is not possible, the results arising from approximate theory serve as useful benchmarks for assessing the efficiency of any design under a particular optimality criterion.

For given p, t, and given a specific optimality criterion Φ, let d^* be a universally optimal and hence Φ-optimal approximate design. Now suppose n is also specified and consider an exact design d with n subjects and the same p, t, as d^*. As before, let C_d and $\bar{C}_d$ denote the information matrices of d for inference on direct and carryover effects, respectively.

In view of Theorem 4.5.1, the Φ-efficiency of d for direct effects, denoted by $E_\Phi(d)$, is defined as

$$E_\Phi(d) = \Phi(C_d)/\{nb(t-1)^{-1}\Phi(H_t)\}, \tag{4.8.1}$$

where b is given by (4.4.3) if $p > t$ and by (4.4.11) if $p \le t$. Similarly, the Φ-efficiency of d for carryover effects, denoted by $\bar{E}_\Phi(d)$, is defined as

$$\bar{E}_\Phi(d) = \Phi(\bar{C}_d)/\{n\bar{b}(t-1)^{-1}\Phi(H_t)\}, \tag{4.8.2}$$

where $\bar{b}$ is as given in Section 4.7. Indeed, (4.8.1) and (4.8.2) are conservative in the sense that they are relative to the approximate optimal design d^*, which may not translate itself to an exact design for the specified n. In this sense, (4.8.1) and (4.8.2) are actually lower bounds on efficiency as long as one restricts oneself to exact designs.

We now use (4.8.1) and (4.8.2) to study the designs constructed by Patterson (1952) which are useful from a practical point of view as they need a small number of subjects with $p \leq t$. In Section 3.4, these designs were shown to be universally optimal for direct and carryover effects in the restricted class of designs where no treatment follows itself. It is of interest to examine how efficient these designs are in the entire class $\Omega_{t,n,p}$. As noted in the proof of Theorem 3.4.2, for a Patterson design d,

$$C_d = \frac{n(p-1)}{t-1}\left[1 - \frac{t}{p(pt-t-1)}\right] H_t.$$

Similarly, from (3.4.5), one can check that for such a design,

$$\bar{C}_d = \frac{n(p-1)}{(t-1)p^2t}\left[p(pt-t-1)-t\right] H_t.$$

Hence, using (4.8.1) and (4.8.2), it can be shown on simplification that the efficiency of a Patterson design for the estimation of direct and carryover effects are, respectively, given by

$$E_\Phi(d) = \frac{U_1}{U_1+t}, \quad \bar{E}_\Phi(d) = \frac{U_2}{U_2+t}, \tag{4.8.3}$$

for any strictly concave Φ satisfying $\Phi(kH_t) = k\Phi(H_t)$ for every $k > 0$. In (4.8.3),

$$U_1 = t^2(p-1)^2(pt(p-1)-p-t),$$
$$U_2 = p(pt-t-1)-t.$$

In Chapter 3, Table 3.5.1 gives a set of values of t, p and n for which Patterson designs are available. Table 4.8.1 shows $E_\Phi(d)$ and $\bar{E}_\Phi(d)$ for these designs, as obtained via (4.8.3).

When $p \leq t$, Kushner (1998) showed that no design can be universally optimal for both direct and carryover effects in the general class while Table 4.8.1 shows that Patterson designs have high efficiencies for estimating both direct and carryover effects in the general class.

In a similar way the efficiencies of other designs may be computed, e.g., the designs of Stufken (1991) (see Theorem 3.2.4), for the case where n is not a multiple of $(p-1)t$, can be seen to have high efficiencies.

TABLE 4.8.1
Efficiency Lower Bounds for Patterson Designs

p	t	n	$E_{\Phi}(d)$	$\bar{E}_{\Phi}(d)$
3	3	6	0.993103	0.800000
3	7	21	0.998885	0.820513
3	8	56	0.999156	0.822222
3	11	55	0.999563	0.825397
4	4	4	0.999306	0.909091
4	5	20	0.999564	0.910714
4	7	14	0.999783	0.912500
4	8	56	0.999835	0.913043
4	13	52	0.999939	0.914474
5	5	10	0.999861	0.947368
5	7	21	0.999930	0.948148
5	8	56	0.999947	0.948387
5	11	55	0.999972	0.948837
5	13	39	0.999980	0.949020
6	6	6	0.999960	0.965517
6	7	42	0.999971	0.965686
6	8	56	0.999978	0.965812
6	11	22	0.999988	0.966049

Chapter 5

Optimality under Some Other Additive Models

5.1 Introduction

In Chapters 2–4, we have presented various optimality results for crossover designs under the traditional model (1.2.1) or its version with correlated errors, i.e., (1.3.14). These models assume that the carryover effect of a treatment remains the same no matter which treatment is applied in the following period and moreover, that it has no relationship with the direct effect. However, there may be several practical situations where such an assumption is untenable. For more on criticisms of model (1.2.1), specially for medical applications, see, e.g., Senn (1992) and Matthews (1994). For examples in other areas, see Kempton, Ferris and David (2001). For these situations it is believed that it might be more reasonable to model the carryover effects in a way different from what has been done in the traditional model. In this chapter, we consider two such variants of model (1.2.1) that have been considered in the literature and review optimality results under them.

The first of these variants assumes that the carryover effect of each treatment is of two types, depending on whether the treatment is followed by itself or by any other treatment. Thus the carryover effect of a treatment on itself is different from that on other treatments. Afsarinejad and Hedayat (2002) initiated work under such a model and termed the two types of carryover effects as *self* and *mixed* carryover effects, respectively. They also gave examples of experimental situations where the experimenter would like to study these two different carryover effects. Their work, however, relates to only two-period crossover designs. Kunert and Stufken (2002) also considered this kind of model and obtained optimal designs for $p \geq 3$. It is interesting to note that in contrast to the optimal designs under (1.2.1) which often have the same treatment applied to the same subject in consecutive periods (e.g., the strongly balanced designs considered in Chapter 2), the optimal designs under a model with self and mixed carryover effects avoid this. Hedayat and Yan (2008) considered a version of this model with correlated errors and their results will be discussed later in Section 7.3.

The second modification of the model (1.2.1) that we consider in this chapter is one in which the carryover effect of a treatment is assumed to be proportional to its direct effect. The optimality aspects of designs under this modified model were considered by Kempton *et al.* (2001), Bailey and Kunert (2006) and Bose and Stufken (2007). This model is conceptually quite simple — a treatment with a larger direct effect will tend to have a larger lingering effect. It is, however, technically more complicated, being intrinsically nonlinear in nature.

In Section 5.2 a model with self and mixed carryover effects is considered and results on optimal designs under such a model are reviewed. In Section 5.3, we consider the determination of optimal designs under a model in which the carryover effects are proportional to the direct effects. Some examples of optimal designs under the respective models are also given in both sections.

5.2 A Model with Self and Mixed Carryover Effects

Kunert and Stufken (2002) gave an interesting example of application of crossover designs in sensory evaluation trials to motivate the model with self and mixed carryover effects. Suppose an assessor is examining several products (for example, tasting the bitterness of different brands of beer) in a sequence. If an assessed product is very bitter, then experience shows that assessors tend to rate the immediately next assessed product (different from the very bitter product) with a lower than normal value of bitterness. If an assessor gets this very bitter product twice in consecutive periods, then usually the same rating is given. Thus the carryover effect of the product is different in the two cases, depending on whether the same or a different product is assessed in two successive periods. A similar behavior of the carryover effects can be observed in other situations too.

In view of this, for such situations, Afsarinejad and Hedayat (2002) proposed a modificaion of the model (1.2.1) and introduced the concepts of self and mixed carryover effects. This model is given by

$$Y_{ij} = \begin{cases} \alpha_i + \beta_j + \tau_{d(i,j)} + \nu_{d(i-1,j)} + \epsilon_{ij}, \text{ if } d(i,j) \neq d(i-1,j), \\ \alpha_i + \beta_j + \tau_{d(i,j)} + \chi_{d(i-1,j)} + \epsilon_{ij}, \text{ if } d(i,j) = d(i-1,j), \end{cases} \quad (5.2.1)$$
$$1 \leq i \leq p, \ 1 \leq j \leq n.$$

where χ_s is the self carryover effect and ν_s is the mixed carryover effect of treatment s, $\nu_{d(0,j)} = \chi_{d(0,j)} = 0$, and all other terms in (5.2.1) have their usual meanings as in (1.2.1), the errors being uncorrelated. Thus (5.2.1) and (1.2.1) differ in their assumptions on the nature of carryover effects.

Under the model (5.2.1), Afsarinejad and Hedayat (2002) studied designs with only two periods. For proving their results they used the technique of Hedayat and Zhao (1990) described in Section 3.3 where a connection between optimal two-period crossover designs and optimal block designs was established. For example, using this technique, Afsarinejad and Hedayat (2002) proved that any symmetric BIB design with t blocks of size k can be used to construct a design which is optimal for direct effects over the subclass of $\Omega_{t,tk,2}$ consisting of designs which are uniform on the first period.

In their method of construction, the t blocks of the symmetric BIB design are first written out as a $k \times t$ array, forming a Youden design (this is always possible). One row is added to this Youden design to form a $(k+1) \times t$ Latin rectangle A, i.e., no treatment appears more than once in a column of A and each treatment appears exactly once in each row of A. Again, this is always possible. Now, the two-period crossover design is constructed by allocating treatment s to the first period of any k subjects, and assigning in the second period of these k subjects, the k treatments which appear in the column of A which has s in row $(k+1)$, $1 \leq s \leq t$. Then, as in Corollary 3.3.1, the optimality of this design for direct effects follows.

We illustrate their method with an example of an optimal design given by them. For more examples and other details, we refer to the original source.

Example 5.2.1. Let $t = 7$. Consider the following BIB design with seven blocks of size $k = 3$:

$$\begin{array}{rr} \text{Block I}: \ 1\ 2\ 4 & \text{Block II}: 2\ 3\ 5 \\ \text{Block III}: 3\ 4\ 6 & \text{Block IV}: 4\ 5\ 7 \\ \text{Block V}: \ 5\ 6\ 1 & \text{Block VI}: 6\ 7\ 2 \\ \text{Block VII}: 7\ 1\ 3 & \end{array}$$

This BIB design leads to the following two-period design which is universally optimal for the estimation of direct effects under the model (5.2.1) over a subclass of $\Omega_{7,21,2}$ consisting of designs which are uniform on the first period.

$$\begin{array}{ccccccc} 3\,3\,3 & 4\,4\,4 & 5\,5\,5 & 6\,6\,6 & 7\,7\,7 & 1\,1\,1 & 2\,2\,2 \\ 1\,2\,4 & 2\,3\,5 & 3\,4\,6 & 4\,5\,7 & 5\,6\,1 & 6\,7\,2 & 7\,1\,3 \end{array}.$$

□

Kunert and Stufken (2002) considered the case $p > 2$, $t > 2$ and obtained optimal designs under model (5.2.1). For proving their results they used a generalization of Lemma 1.4.2 and some techniques developed by Kunert and Martin (2000a) for a general interference model. The results presented in the rest of this section are due to Kunert and Stufken (2002).

As in (1.3.3), model (5.2.1) can be rewritten in matrix notation as

$$\boldsymbol{Y}_d = P\boldsymbol{\alpha} + U\boldsymbol{\beta} + T_d\boldsymbol{\tau} + G_d\boldsymbol{\nu} + S_d\boldsymbol{\chi} + \boldsymbol{\epsilon}, \tag{5.2.2}$$

where $\boldsymbol{\chi} = (\chi_1, \ldots \chi_t)'$, $\boldsymbol{\nu} = (\nu_1, \ldots, \nu_t)'$, and G_d and S_d are the design matrices for the mixed carryover and self carryover effects, respectively, all other notations being as defined in the context of model (1.3.3).

Remark 5.2.1. If a design d is balanced, then obviously, there are no self carryover effects and consequently, $S_d = \mathbf{0}$. For such a design, model (5.2.2) reduces to the traditional model (1.3.3) and thus, as observed in (1.3.5),

$$T_d'G_d = Z_d = \frac{n(p-1)}{t(t-1)}\left(J_t - I_t\right).$$

□

As in (1.3.13), under model (5.2.2), the information matrix for the estimation of direct effects under the design d is given by

$$C_d = T_d'\mathrm{pr}^{\perp}\left([P\ U\ G_d\ S_d]\right)T_d. \tag{5.2.3}$$

In view of the facts that $P\mathbf{1}_p = U\mathbf{1}_n = \mathbf{1}_{np}$ and $T_d\mathbf{1}_t = \mathbf{1}_{np}$ (see (1.3.5)), $\mathbf{1}_{np}$ belongs to the column space of $[P\ U\ G_d\ S_d]$ and the information matrix in (5.2.3) has row and column sums zero for any design $d \in \Omega_{t,n,p}$. Instead of studying C_d directly, Kunert and Stufken (2002) used a slightly simpler matrix which majorizes C_d in the Loewner sense. This helped them to obtain an upper bound for the trace of C_d, $d \in \Omega_{t,n,p}$, and then use Theorem 1.4.1 to find an optimal design. We first state the following result which can be obtained as in (1.4.6) and (1.4.7).

Lemma 5.2.1. *For a design $d \in \Omega_{t,n,p}$,*

$$C_d \leq T_d' \mathrm{pr}^{\perp}([U\ G_d\ S_d])T_d$$

in the Loewner sense, with equality if and only if

$$T_d' \mathrm{pr}^{\perp}([U\ G_d\ S_d])P = \mathbf{0}. \tag{5.2.4}$$

□

We now identify designs for which (5.2.4) holds. For $d \in \Omega_{t,n,p}$, let

$$\begin{aligned}
A_{d11} &= T_d' \mathrm{pr}^{\perp}(U)T_d, \\
A_{d12} &= T_d' \mathrm{pr}^{\perp}(U)G_d, \\
A_{d13} &= T_d' \mathrm{pr}^{\perp}(U)S_d, \\
A_{d22} &= G_d' \mathrm{pr}^{\perp}(U)G_d, \\
A_{d23} &= G_d' \mathrm{pr}^{\perp}(U)S_d, \\
A_{d33} &= S_d' \mathrm{pr}^{\perp}(U)S_d.
\end{aligned}$$

We then have the following result.

Theorem 5.2.1. *Consider a design $d \in \Omega_{t,n,p}$ which satisfies the following conditions:*
(i) each of A_{duv}, $1 \leq u \leq v \leq 3$, is completely symmetric,
(ii) d is uniform on periods,
(iii) the mixed carryover effects of all treatments occur equally often in each period, and
(iv) the self carryover effects of all treatments occur equally often in each period.
Then for such a design d, $T_d' \mathrm{pr}^{\perp}([U\ G_d\ S_d])P = \mathbf{0}$.

Proof. Using (1.3.5), we have

$$\begin{aligned}
\mathbf{1}_t' T_d' \mathrm{pr}^{\perp}(U) &= \mathbf{1}_t'(T_d' - p^{-1}T_d'UU') \\
&= \mathbf{1}_{np}' - p^{-1}\mathbf{1}_{np}'UU' \\
&= \mathbf{1}_{np}' - \mathbf{1}_{np}' = \mathbf{0}' \\
\Rightarrow \quad \mathbf{1}_t' A_{d11} &= \mathbf{0}'.
\end{aligned} \tag{5.2.5}$$

Similarly, one can show that

$$\mathbf{1}_t' A_{d12} = \mathbf{0}' = \mathbf{1}_t' A_{d13}.$$

Therefore, by condition (i), as applied to A_{d11}, A_{d12} and A_{d13}, each of these matrices is a multiple of $H_t = I_t - t^{-1}J_t$. Furthermore, by the same condition, every other A_{duv} admits a generalized inverse which is a linear combination of H_t and $I_t - H_t(= t^{-1}J_t)$. Using these facts, one can check that

$$\begin{aligned} T_d'\mathrm{pr}^{\perp}([U\ G_d\ S_d])P &= T_d'\mathrm{pr}^{\perp}(U)P - A_{d12}A_{d22}^{-}G_d'\mathrm{pr}^{\perp}(U)P \\ &\quad -(A_{d13} - A_{d12}A_{d22}^{-}A_{d23})(A_{d33} - A_{d23}'A_{d22}^{-}A_{d23})^{-} \\ &\quad \times(G_d'\mathrm{pr}^{\perp}(U)P - A_{d23}'A_{d22}^{-}S_d'\mathrm{pr}^{\perp}(U)P) \\ &= T_d'\mathrm{pr}^{\perp}(U)P - x_1H_tG_d'\mathrm{pr}^{\perp}(U)P \\ &\quad -x_2H_t(G_d'\mathrm{pr}^{\perp}(U)P - x_3H_tS_d'\mathrm{pr}^{\perp}(U)P), \end{aligned} \tag{5.2.6}$$

for some numbers x_1, x_2 and x_3. In the above, $\times$ stands for usual matrix multiplication.

Now, by condition (iv) of the theorem, each column of $S_d'(\mathbf{1}_n \otimes I_p)$ is a multiple of $\mathbf{1}_t$. Also, recalling from Section 1.3 that $P = \mathbf{1}_n \otimes I_p$ and $U = I_n \otimes \mathbf{1}_p$, one gets $\mathrm{pr}^{\perp}(U)P = \mathbf{1}_n \otimes H_p$. Therefore, each column of $S_d'\mathrm{pr}^{\perp}(U)P$ is a multiple of $\mathbf{1}_t$ because

$$S_d'\mathrm{pr}^{\perp}(U)P = S_d'(\mathbf{1}_n \otimes H_p) = S_d'(\mathbf{1}_n \otimes I_p)H_p.$$

In a similar manner, using conditions (ii) and (iii), one can see that each column of $T_d'\mathrm{pr}^{\perp}(U)P$ and $G_d'\mathrm{pr}^{\perp}(U)P$ is also a multiple of $\mathbf{1}_t$. Thus

$$H_tG_d'\mathrm{pr}^{\perp}(U)P = H_tS_d'\mathrm{pr}^{\perp}(U)P = \mathbf{0}$$

and invoking (5.2.5) we get

$$T_d'\mathrm{pr}^{\perp}(U)P = \mathbf{0}.$$

Hence from (5.2.6) it is clear that $T_d'\mathrm{pr}^{\perp}([U\ G_d\ S_d])P = \mathbf{0}$ and this completes the proof. □

By Lemma 5.2.1, for a design d satisfying the conditions of Theorem 5.2.1, we have

$$C_d = T_d'\mathrm{pr}^{\perp}([U\ G_d\ S_d])T_d.$$

The next step is to identify designs which satisfy the conditions of Theorem 5.2.1. Analogous to the definitions given in Section 2.5, we now have the following definitions. As usual, the subjects are taken as blocks and for $z > 0$, $[z]$ denotes the largest integer not exceeding z.

Definition 5.2.1. A design $d \in \Omega_{t,n,p}$ is called a balanced block design in the direct effects if

(a) every treatment appears equally often in d,
(b) in each subject, every treatment is allocated either $[p/t]$ or $[p/t] + 1$ times and,

(c) the number of subjects where treatments s and s' are both allocated $[p/t]+1$ times is the same for every $s \neq s'$, $1 \leq s, s' \leq t$.

If $t|p$, then a balanced block design in the direct effects allocates every treatment p/t times to each subject and thus the design is also uniform on subjects (see Definition 2.2.2).

Definition 5.2.2. A design $d \in \Omega_{t,n,p}$ is called a balanced block design in the carryover effects if the first $p-1$ periods of d form a balanced block design in the direct effects in $\Omega_{t,n,p-1}$.

Remark 5.2.2. Clearly, if $d \in \Omega_{t,n,p}$ is a balanced block design in the direct effects, then A_{d11} is completely symmetric. Again, let d be a design for which $z_{dss} = 0$, $1 \leq s \leq t$ (e.g., d could be a balanced design). Then $S_d = \mathbf{0}$ as observed in Remark 5.2.1, implying that $A_{d13} = \mathbf{0} = A_{d23} = A_{d33}$ and thus trivially, each of these matrices is completely symmetric. If this design d with $z_{dss} = 0$ is, in addition, a balanced block design in the carryover effects, then A_{d22} is also completely symmetric. Finally, if d is balanced and $T_d'UU'G_d$ is completely symmetric, then from the form of $T_d'G_d$ given in Remark 5.2.1, it is clear that A_{d12} is completely symmetric. □

We now define a class of crossover designs for which (5.2.4) will be shown to hold.

Definition 5.2.3. A design $d \in \Omega_{t,n,p}$ is called totally balanced if
(a) d is uniform on periods,
(b) d is balanced (in the sense of Definition 2.2.4),
(c) d is a balanced block design in the direct effects,
(d) d is a balanced block design in the carryover effects, and
(e) the number of subjects where both treatments s and s' appear $[p/t]+1$ times and treatment s' does not appear in the last period is the same for every pair s, s', $1 \leq s, s' \leq t$; $s \neq s'$.

Clearly, a totally balanced design may have $p < t$, $p = t$ or $p > t$, and when $t|p$, it is simply a balanced uniform design. Some examples of totally balanced designs are given in the next section. The following result gives the information matrix for direct effects for these designs.

Theorem 5.2.2. *For a totally balanced design $d \in \Omega_{t,n,p}$,*

$$C_d = T_d' \mathrm{pr}^{\perp}([U\ G_d\ S_d])T_d.$$

Proof. By Lemma 5.2.1, it suffices to show that the conditions of Theorem 5.2.1 are satisfied by a totally balanced design d. Since d is uniform on periods, it meets condition (ii) of Theorem 5.2.1. Again, since d is balanced, condition (iii) is satisfied and condition (iv) is trivially satisfied since in a balanced design there are no self carryover effects. In order to complete the proof, we now need to verify condition (i) for d. It is evident from Remark 5.2.2 that for a totally balanced design d, $A_{d11}, A_{d13}, A_{d22}, A_{d23}$ and A_{d33} are all completely symmetric. It remains to show that A_{d12} is also completely symmetric. For this, we separately study two cases, depending on whether t divides p or not.

Case 1: t divides p.

In this case, d is uniform on subjects and is actually a generalized Latin square design (see Definition 2.5.4) for which condition (e) of Definition 5.2.3 is trivially true. In d, a treatment s appears in the last period in n/t subjects and for these subjects, the mixed carryover effect of s appears $p/t - 1$ times. For all other (i.e., $n - n/t$) subjects, the mixed carryover effect of s appears p/t times. Thus

$$\mathbf{1}'_{np} G_d = \left\{ \left(\frac{p}{t} - 1 \right) \frac{n}{t} + \frac{p}{t} \left(n - \frac{n}{t} \right) \right\} \mathbf{1}'_t.$$

The above, together with the fact that $T'_d U = N_d = pt^{-1} J_{tn}$ for this design d, implies that

$$\begin{aligned} T'_d U U' G_d = &\quad pt^{-1} J_{tn} U' G_d = pt^{-1} \mathbf{1}_t \mathbf{1}'_{np} G_d \\ &= \left\{ \left(\frac{p}{t} - 1 \right) \frac{np}{t^2} + \frac{p^2}{t^2} \left(n - \frac{n}{t} \right) \right\} J_t, \end{aligned}$$

which is completely symmetric. Recalling Remark 5.2.2, it is now clear that A_{d12} is also completely symmetric.

Case 2: t does not divide p.

Note that d is balanced in the carryover effects. Since $[(p-1)/t] = [p/t]$, it now follows that in d, the mixed carryover effect of each treatment appears in each subject either $[p/t]$ times or $[p/t] + 1$ times. So, a treatment that appears only $[p/t]$ times in a subject cannot appear in the last period of this subject. For treatments s and s', $1 \le s, s' \le t; s \ne s'$, let u_1, u_2 and u_3 be defined as follows:

(i) u_1 is the number of subjects where treatments s and s' both appear $[p/t] + 1$ times and treatment s' does not appear in the last period,

(ii) u_2 is the number of subjects where treatment s appears $[p/t] + 1$ times and the mixed carryover effect of s' appears $[p/t]$ times, and

(iii) u_3 is the number of subjects where treatment s appears $[p/t]$ times and the mixed carryover effect of treatment s' appears $[p/t]+1$ times.

Then one can check that, for $s \neq s'$, the (s,s')th element of $T_d'UU'G_d$ equals

$$(n-u_1-u_2-u_3)[p/t]^2 + (u_2+u_3)[p/t]\left([p/t]+1\right) + u_1([p/t]+1)^2.$$

Since d satisfies condition (e) of Definition 5.2.3, u_1 is a constant for all $s \neq s'$. Again, the number of subjects in d where s appears $[p/t]+1$ times is independent of s and this number is u_1+u_2, implying that u_2 does not depend on s or s'. Similarly, the number of subjects where the mixed carryover effect of treatment s' appears $[p/t]+1$ times is also independent of s' and this number is u_1+u_3. Hence u_3 too does not depend on s or s' and all off-diagonal elements of $T_d'UU'G_d$ are the same. So, noting the complete symmetry of $T_d'G_d$, (see Remark 5.2.1), it is clear that $A_{d12} = T_d'G_d - p^{-1}T_d'UU'G_d$ also has all off-diagonal elements equal. Now, recalling from (5.2.5) that $\mathbf{1}_t'A_{d12} = \mathbf{0}'$ for all $d \in \Omega_{t,n,p}$, for a totally balanced design, the diagonal elements of A_{d12} are also the same. Hence A_{d12} is a completely symmetric matrix. This completes the proof. □

Once an upper bound on C_d has been found and designs attaining the bound are identified, a key step in obtaining universally optimal designs involves maximization of the trace of this upper bound. To that end, we first define for any design $d \in \Omega_{t,n,p}$,

$$a_{duv} = \mathrm{tr}(H_tA_{duv}H_t) = \mathrm{tr}(H_tA_{duv}),\ 1 \le u \le v \le 3. \tag{5.2.7}$$

Kunert and Stufken (2002) presented the following result.

Lemma 5.2.2. *For every $d \in \Omega_{t,n,p}$,*

$$\mathrm{tr}\left(T_d'\mathrm{pr}^{\perp}([U\ G_d\ S_d])T_d\right) \le q_d^* \tag{5.2.8}$$

where q_d^* is defined in terms of the values of a_{duv} as follows:

1. *If $a_{d22}a_{d33} - a_{d23}^2 > 0$, then*

$$q_d^* = a_{d11} - \frac{a_{d12}^2a_{d33} - 2a_{d12}a_{d13}a_{d23} + a_{d13}^2a_{d22}}{a_{d22}a_{d33} - a_{d23}^2}.$$

2. *If $a_{d22}a_{d33} - a_{d23}^2 = 0$ and $a_{d22} > 0$, then $q_d^* = a_{d11} - a_{d12}^2/a_{d22}$.*
3. *If $a_{d22} = 0$ and $a_{d33} > 0$, then $q_d^* = a_{d11} - a_{d13}^2/a_{d33}$.*
4. *If $a_{d22} = a_{d33} = 0$, then $q_d^* = a_{d11}$.*

Furthermore, equality in (5.2.8) holds if all A_{duv}, $1 \le u \le v \le 3$, are completely symmetric. □

As noted in the proof of Theorem 5.2.2, the matrices A_{duv} are all completely symmetric for a totally balanced design. Hence Lemmas 5.2.1, 5.2.2 and Theorem 5.2.2 together lead to

Theorem 5.2.3. *For every $d \in \Omega_{t,n,p}$,*

$$\mathrm{tr}(C_d) \le q_d^*,$$

with equality if d is a totally balanced design. □

We now look for a design d for which $\mathrm{tr}(C_d)$ equals the maximum possible value of q_d^*. For obtaining such a design, symmetry blocks or classes of equivalent sequences (*cf.* Definition 4.2.1) as used in Chapter 4 are found useful. For given p and t, suppose $\mathcal{S}$, the set of all t^p sequences, is the union of K distinct symmetry blocks, and as in Chapter 4, let $< \boldsymbol{s} >$ denote the symmetry block or equivalence class containing the sequence $\boldsymbol{s}$. For a design $d \in \Omega_{t,n,p}$, $P_{\boldsymbol{s}}$ is the proportion of subjects to which d assigns sequences from the equivalence class $< \boldsymbol{s} >$, and we write $< \boldsymbol{s} > \in \mathcal{S}$ to denote that $< \boldsymbol{s} >$ runs over all the K equivalence classes in $\mathcal{S}$.

As these sequences are applied to individual subjects of d, the contribution of the different subjects to $\mathrm{tr}(C_d)$ needs to be evaluated. Let T_{dj}, G_{dj} and S_{dj}, respectively, denote the design matrices for the direct effects, mixed carryover effects and self carryover effects with respect to the subject j, $1 \le j \le n$. For $1 \le j \le n$, define

$$\begin{aligned}
a_{d11}^{(j)} &= \mathrm{tr}(H_t(T'_{dj}T_{dj} - p^{-1}T'_{dj}J_pT_{dj})),\\
a_{d12}^{(j)} &= \mathrm{tr}(H_t(T'_{dj}G_{dj} - p^{-1}T'_{dj}J_pG_{dj})),\\
a_{d13}^{(j)} &= \mathrm{tr}(H_t(T'_{dj}S_{dj} - p^{-1}T'_{dj}J_pS_{dj})),\\
a_{d22}^{(j)} &= \mathrm{tr}(H_t(G'_{dj}G_{dj} - p^{-1}G'_{dj}J_pG_{dj})),\\
a_{d23}^{(j)} &= \mathrm{tr}(H_t(G'_{dj}S_{dj} - p^{-1}G'_{dj}J_pS_{dj})),\\
a_{d33}^{(j)} &= \mathrm{tr}(H_t(S'_{dj}S_{dj} - p^{-1}S'_{dj}J_pS_{dj})),
\end{aligned}$$

so that from (5.2.7) and the fact that $U = I_n \otimes \mathbf{1}_p$,

$$a_{duv} = \sum_{j=1}^{n} a_{duv}^{(j)}, \ 1 \le u \le v \le 3.$$

Clearly, the value of $a_{duv}^{(j)}$ depends on the treatment sequence applied to subject j and it is the same for all subjects which receive sequences from the same equivalence class. Let

$$a_{uv}(\boldsymbol{s}) = a_{duv}^{(j_{\boldsymbol{s}})}, \ 1 \le u \le v \le 3,$$

where $j_{\boldsymbol{s}}$ is any subject receiving a sequence from $< \boldsymbol{s} >$. Then

$$a_{duv} = n \left(\sum_{<\boldsymbol{s}> \in \mathcal{S}} P_{\boldsymbol{s}} a_{uv}(\boldsymbol{s}) \right), \ 1 \le u \le v \le 3. \tag{5.2.9}$$

Thus the proportions $P_{\boldsymbol{s}}$ determine q_d^* and so, one needs to choose these proportions in such a manner that q_d^* is maximized. However, from the form of q_d^* as defined in Lemma 5.2.2 and (5.2.9), it is clear that q_d^* is a nonlinear function of the $P_{\boldsymbol{s}}$ and the problem of determining the proportions which maximize q_d^* becomes a challenging one. This problem is overcome by introducing the bivariate quadratic function

$$q_d(x, y) = a_{d11} + 2xa_{d12} + x^2 a_{d22} + 2ya_{d13} + y^2 a_{d33} + 2xya_{d23}. \tag{5.2.10}$$

Again, for a sequence $\boldsymbol{s} \in \mathcal{S}$, define

$$h_{\boldsymbol{s}}(x, y) = a_{11}(\boldsymbol{s}) + 2xa_{12}(\boldsymbol{s}) + x^2 a_{22}(\boldsymbol{s}) + 2ya_{13}(\boldsymbol{s}) + y^2 a_{33}(\boldsymbol{s}) + 2xya_{23}(\boldsymbol{s}). \tag{5.2.11}$$

Clearly, $h_{\boldsymbol{s}}(x, y)$ is the same for all sequences in the same equivalence class, i.e., for all $\boldsymbol{s} \in < \boldsymbol{s} >$ and hence using (5.2.9), (5.2.10) and (5.2.11),

$$q_d(x, y) = n \sum_{<\boldsymbol{s}> \in \mathcal{S}} P_{\boldsymbol{s}} h_{\boldsymbol{s}}(x, y).$$

This makes $q_d(x, y)$ a linear combination of the quantities $\{h_{\boldsymbol{s}}(x, y)\}$.

From (5.2.10) and the expressions for q_d^* shown in Lemma 5.2.2, it can be verified that

$$q_d(x, y) \ge q_d^*, \quad \text{for all } x, y,$$

the lower bound q_d^* being attainable. Furthermore, from the definition of $q_d(x, y)$, it is not hard to see that

$$q_d(x, y) \le n \max_{\boldsymbol{s}} \ h_{\boldsymbol{s}}(x, y), \quad \text{for all } x, y.$$

Hence it follows that for every $d \in \Omega_{t,n,p}$,

$$q_d^* \le n \min_{x,y} \max_{\boldsymbol{s}} \ h_{\boldsymbol{s}}(x, y). \tag{5.2.12}$$

The next two theorems summarize some results of Kunert and Stufken (2002). The proof of the first one of these is quite involved and hence skipped.

Theorem 5.2.4. *For $t \geq 3$ and $3 \leq p \leq 2t$, consider any sequence $\boldsymbol{s}^*$ such that*

(i) no treatment is applied to two consecutive periods in $\boldsymbol{s}^*$,

(ii) the numbers of times any two treatments appear in $\boldsymbol{s}^*$ differs by at most unity, and

(iii) the treatment in the last period appears the maximum number of times in $\boldsymbol{s}^*$.

Then for all x, y,

$$h_{\boldsymbol{s}^*}(x, y) \geq h_{\boldsymbol{s}^*}(x^*, -1),$$

and

$$h_{\boldsymbol{s}^*}(x^*, -1) = \min_{x,y} \max_{\boldsymbol{s}} h_{\boldsymbol{s}}(x, y),$$

where

$$x^* = \begin{cases} \dfrac{t}{(tp - t - 1)}, & \text{for } p \leq t, \\[2ex] \dfrac{tp + 2t(p-1-t)}{pt(t-1) + (pt - 2t - 1)(p - 1 - t)}, & \text{for } p > t. \end{cases}$$

□

If $p \leq t$, then any sequence which satisfies the properties (i)–(iii) in Theorem 5.2.4 must necessarily be equivalent to $(1, 2, \ldots, p)'$. If $p > t$, sequences from more than one single equivalence class can satisfy these conditions. For instance, if $t < p < 2t$, then both the non-equivalent sequences, $(1, 2, \ldots, t, 1, 2, \ldots, p-t)'$ and $(1, 2, \ldots, t, t-1, t-2, \ldots, 2t-p)'$, satisfy the conditions (i)–(iii) in Theorem 5.2.4.

Theorem 5.2.5. *For $t \geq 3$ and $3 \leq p \leq 2t$, if a totally balanced design $d^* \in \Omega_{t,n,p}$ exists then d^* is universally optimal for the estimation of direct effects over $\Omega_{t,n,p}$ under the model (5.2.1).*

Proof. For the design d^*, the information matrix for direct effects under the model (5.2.1) is as given by Theorem 5.2.2, that is,

$$C_{d^*} = T'_{d^*} \mathrm{pr}^{\perp}([U \; G_{d^*} \; S_{d^*}]) T_{d^*}. \tag{5.2.13}$$

In the proof of Theorem 5.2.2, it was argued that each A_{d^*uv}, $1 \leq u \leq v \leq 3$, is completely symmetric. Hence invoking Lemma 1.2.1 with $B = U$, $C = [G_{d^*} \; S_{d^*}]$, and using the fact that $S_{d^*} = 0$, one can check that C_{d^*} as shown in (5.2.13) is completely symmetric.

In view of Theorem 1.4.1, it is now enough to show that d^* maximizes $\mathrm{tr}(C_d)$ over $\Omega_{t,n,p}$. From (5.2.12) and Theorem 5.2.3, first observe that,

$$\mathrm{tr}(C_d) \le q_d^* \le n \min_{x,y} \max_{\boldsymbol{s}} h_{\boldsymbol{s}}(x,y),$$

for every $d \in \Omega_{t,n,p}$, and that

$$\mathrm{tr}(C_{d^*}) = q_{d^*}^*.$$

Hence it would suffice to show that

$$q_{d^*}^* = n \min_{x,y} \max_{\boldsymbol{s}} h_{\boldsymbol{s}}(x,y).$$

To prove the last identity now observe that the conditions (b), (c) and (d) of Definition 5.2.3, respectively, imply the conditions (i), (ii) and (iii) of Theorem 5.2.4. Therefore, every sequence of a totally balanced design d^* meets all the conditions of Theorem 5.2.4. Hence by that theorem,

$$q_{d^*}(x,y) = n \sum_{<\boldsymbol{s}>\in\mathcal{S}} P_{\boldsymbol{s}} h_{\boldsymbol{s}}(x,y) \ge n \min_{x,y} \max_{\boldsymbol{s}} h_{\boldsymbol{s}}(x,y),$$

for every x, y. Since $q_{d^*}^*$ is an attainable lower bound of $q_{d^*}(x,y)$, the above implies that

$$q_{d^*}^* \ge n \min_{x,y} \max_{\boldsymbol{s}} h_{\boldsymbol{s}}(x,y),$$

that is,

$$q_{d^*}^* = n \min_{x,y} \max_{\boldsymbol{s}} h_{\boldsymbol{s}}(x,y),$$

in view of (5.2.12). The result is now immediate. □

Remark 5.2.3. Interestingly, totally balanced designs are also optimal under a model in which the carryover effects are proportional to direct effects. These results are reviewed in the next section. □

Corollary 5.2.1. *If $p = t$ or $p = 2t$ and a generalized Latin square design d^* exists in $\Omega_{t,n,p}$ such that d^* is balanced (in the sense of Definition 2.2.4), then d^* is universally optimal for direct effects over $\Omega_{t,n,p}$ under the model (5.2.1).*

Proof. The result follows from Theorem 5.2.5 on observing that the design d^* satisfies all the conditions of Definition 5.2.3 and is thus totally balanced. □

Note that in a totally balanced design no treatment follows itself on any subject. Therefore, the optimal designs under the model (5.2.1) are in many cases different from those under the traditional model (1.2.1) where strongly balanced designs (see Definition 2.2.5) are often optimal.

We now present some examples of totally balanced designs from Kunert and Stufken (2002). Each of these designs is universally optimal for direct effects over $\Omega_{t,n,p}$ under (5.2.1).

Example 5.2.2. $t = 4, p = 3$. A totally balanced design d^* with $n = 12$ is

$$d^* = \begin{matrix} 1 & 3 & 2 & 4 & 1 & 2 & 1 & 4 & 3 & 4 & 2 & 3 \\ 2 & 1 & 3 & 2 & 4 & 1 & 3 & 1 & 4 & 3 & 4 & 2 \\ 3 & 2 & 1 & 1 & 2 & 4 & 4 & 3 & 1 & 2 & 3 & 4 \end{matrix}. \qquad \square$$

For $t = 4, p = 3$, the design which is optimal for direct effects over $\Omega_{t,n,p}$ under the traditional model (1.2.1) has some treatment sequences equivalent to $(1, 2, 2)'$ (see Example 4.6.2) and is thus different from the above design d^*.

Example 5.2.3. Let $t = 4 = p$. A totally balanced design d^* with $n = 4$ is

$$d^* = \begin{matrix} 1 & 2 & 3 & 4 \\ 2 & 4 & 1 & 3 \\ 3 & 1 & 4 & 2 \\ 4 & 3 & 2 & 1 \end{matrix}.$$

This d^* is a Latin Square Design which is also balanced. This is also universally optimal under the traditional model (1.2.1) (see Example 2.3.1). $\square$

Example 5.2.4. Let $t = 3, p = 4$. A totally balanced design d^* with $n = 6$ is

$$d^* = \begin{matrix} 1 & 2 & 3 & 3 & 1 & 2 \\ 2 & 3 & 1 & 2 & 3 & 1 \\ 3 & 1 & 2 & 1 & 2 & 3 \\ 1 & 2 & 3 & 3 & 1 & 2 \end{matrix}.$$

This d^* is different from the optimal design in $\Omega_{3,6,4}$ under model (1.2.1), namely, the extra-period strongly balanced design (see Example 2.4.2). $\square$

Example 5.2.5. Let $t = 3, p = 6$. A totally balanced design d^* with $n = 6$ is given by

$$d^* = \begin{matrix} 1 & 2 & 3 & 3 & 1 & 2 \\ 2 & 3 & 1 & 1 & 2 & 3 \\ 3 & 1 & 2 & 2 & 3 & 1 \\ 2 & 3 & 1 & 3 & 1 & 2 \\ 1 & 2 & 3 & 2 & 3 & 1 \\ 3 & 1 & 2 & 1 & 2 & 3 \end{matrix}.$$

This d^* is different from the optimal design under model (1.2.1), namely the nearly strongly balanced design (see design d_1 in Example 2.5.3). □

Copies of the designs in Examples 5.2.2–5.2.5 are also totally balanced and hence optimal in $\Omega_{t,n,p}$ where n is a multiple of the number of subjects used in these examples.

In cases where no optimal design exists, one can study the efficiency of an available design. This can be done by comparing the information matrix of an available design with that of a (possibly hypothetical) totally balanced design. For a totally balanced design d^*, Kunert and Stufken (2002) observed that the information matrix for direct effects is given by

$$C_{d^*} = n(t-1)^{-1} h_{\boldsymbol{s}^*}(x^*, -1) H_t,$$

where $h_s(x, y)$ is as defined in (5.2.11), and x^* and $\boldsymbol{s}^*$ are as in Theorem 5.2.4.

5.3 A Model with Carryover Effects Proportional to Direct Effects

In this section, we consider a second variant of the traditional model in which it is assumed that the carryover effect of a treatment is proportional to the direct effect of that treatment.

As stated in Section 5.1, this model is conceptually easy to explain. Kempton *et al.* (2001) gave examples of experiments where such models are found suitable. It is interesting to note that the proportionality constant may be positive or negative, depending on whether the carryover effect is due to assimilation of successive treatments or due to contrasting between successive treatments. An example of the former situation was given by Cross (1973) where the subjects were asked to rate the loudness levels of different sound stimuli and it was found that subjects usually gave a higher rating to a sound stimulus when it was preceded by a loud sound and gave the same sound a lower rating when it was preceded by a soft sound. On the other hand, an example of the second situation was given by Scifferstein and Oudejans (1996) who asked subjects to rate the saltiness of different saline solutions and, the subjects rated a solution to be less salty if immediately preceded by a solution with high salt concentration while they rated the same solution to be more salty when preceded by a low salt solution.

Another advantage of such a model is that it has fewer parameters to estimate since there are no separate carryover effects for treatments.

Technically, however, this model appears to be harder to handle because of its non-linearity.

The issue of identifying optimal designs under such a model has received attention in the recent past. Kempton *et al.* (2001) studied this model for an unknown constant of proportionality. The optimality criterion they used is an average A-optimality criterion; the averaging becomes necessary as the information matrix depends on the true values of the direct effects and the proportionality factor. By using a combination of analytical results and computer search, they obtained designs that are optimal for direct effects for different values of the constant of proportionality. Bailey and Kunert (2006) generalized some of the results of Kempton *et al.* (2001) and developed analytical proofs to show that the totally balanced designs (see Definition 5.2.3) are optimal for direct effects under the average A-optimality criterion. Bose and Stufken (2007) also considered this model but they chiefly studied the case where the constant of proportionality is known and for this simpler situation, they obtained universally optimal designs. In this section, we review some of the results under this model.

Under the assumption that the carryover effect is proportional to the direct effect, the traditional model (1.3.3) is modified to

$$\begin{aligned} \boldsymbol{Y}_d &= \mu \mathbf{1}_{np} + P\boldsymbol{\alpha} + U\boldsymbol{\beta} + T_d\boldsymbol{\tau} + F_d\gamma\boldsymbol{\tau} + \boldsymbol{\epsilon} \\ &= \mu \mathbf{1}_{np} + P\boldsymbol{\alpha} + U\boldsymbol{\beta} + (W_d(\gamma))\boldsymbol{\tau} + \boldsymbol{\epsilon}, \end{aligned} \tag{5.3.1}$$

with $W_d(\gamma) = T_d + F_d\gamma$, γ being the constant of proportionality and all other notation being as in (1.3.3). It is assumed that $\gamma \in [-1, 1]$, an assumption which one might expect to be true in practice, and no other restrictions are placed on γ. When $\gamma = 0$, there are no carryover effects and (5.3.1) reduces to the usual model for row-column designs. We are interested in the estimation of $\boldsymbol{\tau}$ and not γ. Without loss of generality, we may suppose that $\boldsymbol{\tau}'\mathbf{1}_t = 0$. Therefore, if there are only two treatments, each treatment has a separate carryover effect and thus model (5.3.1) becomes equivalent to (1.3.3) when $t = 2$. So, in the subsequent discussion, we study the case $t \geq 3$.

Although model (5.3.1) has fewer parameters than the traditional model (1.3.3), it is more difficult to study (5.3.1) as it is nonlinear in $\boldsymbol{\tau}$ and γ. Kempton *et al.* (2001) used a Taylor series expansion in the neighborhood of the unknown true values of $\boldsymbol{\tau}$ and γ to obtain the following linear approximation to the model (5.3.1):

$$\boldsymbol{Y}_d = \mu \mathbf{1}_{np} + P\boldsymbol{\alpha} + U\boldsymbol{\beta} + (W_d(\gamma_0))\boldsymbol{\tau}_0 + (W_d(\gamma_0))(\boldsymbol{\tau} - \boldsymbol{\tau}_0) + (F_d\boldsymbol{\tau}_0)(\gamma - \gamma_0) + \boldsymbol{\epsilon},$$

or equivalently,

$$\boldsymbol{Y}_d + F_d\boldsymbol{\tau}_0\gamma_0 = \mu\mathbf{1}_{np} + P\boldsymbol{\alpha} + U\boldsymbol{\beta} + (W_d(\gamma_0))\boldsymbol{\tau} + (F_d\boldsymbol{\tau}_0)\gamma + \boldsymbol{\epsilon}, \quad (5.3.2)$$

where $\boldsymbol{\tau}_0$ and γ_0 are the true values of $\boldsymbol{\tau}$ and γ, respectively. The least squares estimator of $(\boldsymbol{\tau}, \gamma)$ under (5.3.1) is asymptotically equivalent to the least squares estimator under (5.3.2); *cf.* Kempton *et al.* (2001) and Bailey and Kunert (2006). As in (1.3.13), for a design $d \in \Omega_{t,n,p}$, the information matrix for estimating $\boldsymbol{\tau}$ under (5.3.2) is given by

$$C_d = (W_d(\gamma_0))'\mathrm{pr}^{\perp}([P\; U\; F_d\boldsymbol{\tau}_0])(W_d(\gamma_0)), \quad (5.3.3)$$

and thus this information matrix depends on the two unknowns $\boldsymbol{\tau}_0$ and γ_0. To overcome this difficulty, Kempton *et al.* (2001) studied the average performance of a design and defined the (i) $\bar{A}$-criterion, where the usual A-criterion is averaged over a distribution of $\boldsymbol{\tau}_0$, and (ii) the IA-criterion, where the A-criterion is averaged over the joint distribution of $\boldsymbol{\tau}_0$ and γ_0. Towards this, for given $\boldsymbol{\tau}_0$ and γ_0, they considered the average variance of comparisons between direct effects as

$$A(\boldsymbol{\tau}_0, \gamma_0) = \sum_{s \neq s'=1}^{t} \frac{\mathrm{Var}(\hat{\boldsymbol{\tau}}_s - \hat{\boldsymbol{\tau}}_{s'})}{t(t-1)},$$

and then defined the $\bar{A}$ criterion as

$$\bar{A}(\gamma_0) = \int A(\boldsymbol{\tau}_0, \gamma_0)dG(\boldsymbol{\tau}_0) \quad (5.3.4)$$

where G represents the multivariate normal distribution for $\boldsymbol{\tau}_0$ with mean zero and dispersion H_t. This is a local criterion as it depends on the unknown γ_0. The global IA-criterion was defined by Kempton *et al.* (2001) as

$$IA = \int\int A(\boldsymbol{\tau}_0, \gamma_0)dG(\boldsymbol{\tau}_0)dF(\gamma_0)$$

where F represents the uniform distribution for γ_0 on $[-1, 1]$.

They proved the following two results. The matrix Z_d in Theorem 5.3.1 is the usual direct-versus-carryover effect incidence matrix, as defined in Chapter 1.

Theorem 5.3.1. *Suppose $n = \mu_1 t$, $\mu_1 \geq 1$ and $p = t$, and let $d \in \Omega_{t,n,p}$ be such that the matrix Z_d is completely symmetric. Then d is $\bar{A}$-optimal for the estimation of direct effects over the class of all uniform designs in $\Omega_{t,\mu_1 t,t}$ for all γ_0.* □

Theorem 5.3.2. *Suppose $n = \mu_1 t$, $\mu_1 \geq 1$ and $p = t+1$, and let $d \in \Omega_{t,n,p}$ be a strongly balanced design which is uniform on periods and uniform on subjects in the first $(p-1)$ periods. Then d is IA-optimal for the estimation of direct effects over $\Omega_{t,n,p}$.* □

Remark 5.3.1. The balanced uniform designs (see Definition 2.2.4) have completely symmetric Z_d matrices and so, by Theorem 5.3.1, these designs are $\bar{A}$-optimal for direct effects under the proportional carryover model; the universal optimality of these designs under the traditional model (1.3.3) has been discussed in Chapter 2. The optimal designs of Theorem 5.3.2 are the extra-period designs which are also universally optimal for direct effects under (1.3.3) (*cf.* Theorem 2.4.2). □

We now discuss the results of Bailey and Kunert (2006) who adopted model (5.3.1) and its linearized version (5.3.2). They showed that the totally balanced designs which are universally optimal under (5.2.1) (*cf.* Theorem 5.2.5) also have optimality properties under this model. We only sketch the proofs, omitting the technical details.

Since $T_d\mathbf{1}_t$ and $F_d\mathbf{1}_t$ are in the column space of P and $\tau_0'\mathbf{1}_t = 0$, it follows from (5.3.3) that

$$C_d = H_t(W_d(\gamma_0))'\mathrm{pr}^{\perp}([P\ U\ F_dH_t\boldsymbol{\tau}_0])(W_d(\gamma_0))H_t.$$

As in Lemma 5.2.1, here too, first a simpler matrix is found which majorizes C_d in the Lowener sense. Let $\tilde{C}_d$ be the information matrix for direct effects if there are no period effects in the model. Then

$$\tilde{C}_d = H_t(W_d(\gamma_0))'\mathrm{pr}^{\perp}([U\ F_dH_t\boldsymbol{\tau}_0])(W_d(\gamma_0))H_t,$$

and hence as in (1.4.6),

$$C_d \leq \tilde{C}_d \tag{5.3.5}$$

with equality holding in (5.3.5) if and only if

$$H_t(W_d(\gamma_0))'\mathrm{pr}^{\perp}([U\ F_dH_t\boldsymbol{\tau}_0])P = \mathbf{0}. \tag{5.3.6}$$

Let

$$\begin{aligned} B_{d11} &= H_tT_d'\mathrm{pr}^{\perp}(U)T_dH_t, \\ B_{d12} &= H_tT_d'\mathrm{pr}^{\perp}(U)F_dH_t, \\ B_{d22} &= H_tF_d'\mathrm{pr}^{\perp}(U)F_dH_t, \text{ and} \\ B_d &= B_{d11} + \gamma_0(B_{d12} + B_{d12}') + \gamma_0^2B_{d22}. \end{aligned} \tag{5.3.7}$$

Using Lemma 1.2.1, we can now express (5.3.5) as

$$C_d \leq B_d - (B_{d12}\boldsymbol{\tau}_0 + \gamma_0B_{d22}\boldsymbol{\tau}_0)(\boldsymbol{\tau}_0'B_{d22}\boldsymbol{\tau}_0)^{-1}(B_{d12}\boldsymbol{\tau}_0 + \gamma_0B_{d22}\boldsymbol{\tau}_0)',$$

and similarly, (5.3.6) can be expressed as

$$H_tT_d'\mathrm{pr}^{\perp}(U)P + \gamma_0 H_tF_d'\mathrm{pr}^{\perp}(U)P \\ -(B_{d12}\boldsymbol{\tau}_0 + \gamma_0 B_{d22}\boldsymbol{\tau}_0)(\boldsymbol{\tau}_0'B_{d22}\boldsymbol{\tau}_0)^{-1}\boldsymbol{\tau}_0'H_tF_d'\mathrm{pr}^{\perp}(U)P = \mathbf{0}. \tag{5.3.8}$$

If a design is uniform on periods, then as in the proof of Theorem 5.2.1, each column of $T_d'\mathrm{pr}^{\perp}(U)P$ and $F_d'\mathrm{pr}^{\perp}(U)P$ is a multiple of $\mathbf{1}_t$ and so

$$H_tT_d'\mathrm{pr}^{\perp}(U)P = H_tF_d'\mathrm{pr}^{\perp}(U)P = \mathbf{0}.$$

Consequently, (5.3.8) holds and hence (5.3.6) holds. Thus $C_d = \tilde{C}_d$ when d is uniform on periods. However, as remarked earlier, the information matrix depends on the unknown $\boldsymbol{\tau}_0$ and γ_0 and so, Bailey and Kunert (2006) used the following criterion.

For any fixed $\boldsymbol{\tau}_0$ and any connected design d in $\Omega_{t,n,p}$, let $\theta_{1d} \geq \theta_{2d} \geq \cdots \geq \theta_{t-1,d}$ be the nonzero eigenvalues of C_d. Define

$$\phi_A(C_d, \boldsymbol{\tau}_0) = \sum_{k=1}^{t-1} \frac{1}{\theta_{kd}}.$$

Bailey and Kunert (2006) assumed that the distribution, $\mathcal{P}$, of $\boldsymbol{\tau}_0$ is permutation invariant, an assumption that will be valid if the treatments are randomized. With such a $\mathcal{P}$ and for any design $d \in \Omega_{t,n,p}$, let

$$\bar{\phi}_A(C_d) = \int \phi_A(C_d, \boldsymbol{\tau}_0)d\mathcal{P}(\boldsymbol{\tau}_0).$$

Then a design which minimizes $\bar{\phi}_A(C_d)$ in a class $\mathcal{D} \subset \Omega_{t,n,p}$ is said to be $\bar{\phi}_A$-optimal for the direct effects over $\mathcal{D}$.

Clearly,

$$\bar{\phi}_A(C_d) = \bar{A}\left(\frac{t-1}{2\sigma^2}\right),$$

where $\bar{A}$ is the $\bar{A}$-criterion as defined in (5.3.4) with G there replaced by $\mathcal{P}$. Since $\boldsymbol{\tau}_0'\mathbf{1}_t = \mathbf{0}$, there exists an orthonormal basis of $\mathbb{R}^t$ of the following form:

$$\left\{\boldsymbol{x}_1, \ldots, \boldsymbol{x}_{t-2}, \boldsymbol{x}_{t-1} = \frac{\boldsymbol{\tau}_0}{\sqrt{\boldsymbol{\tau}_0'\boldsymbol{\tau}_0}}, \boldsymbol{x}_t = \frac{\mathbf{1}_t}{\sqrt{t}}\right\}.$$

Then, exploiting the fact that $C_d \leq \tilde{C}_d$, Bailey and Kunert (2006) showed that

$$\phi_A(C_d, \boldsymbol{\tau}_0) \geq \sum_{k=1}^{t-1} \frac{1}{\boldsymbol{x}_l'\tilde{C}_d\boldsymbol{x}_l} \geq \frac{1}{l_{\boldsymbol{\tau}_0,d}} + \frac{(t-2)^2}{s_{\boldsymbol{\tau}_0,d}}, \tag{5.3.9}$$

where

$$l_{\boldsymbol{\tau}_0,d} = \boldsymbol{x}'_{t-1}\tilde{C}_d\boldsymbol{x}_{t-1}$$
$$= (\boldsymbol{\tau}'_0\boldsymbol{\tau}_0)^{-1}\left\{\boldsymbol{\tau}'_0 B_{d11}\boldsymbol{\tau}_0 - \frac{(\boldsymbol{\tau}'_0 B_{d12}\boldsymbol{\tau}_0)^2}{\boldsymbol{\tau}'_0 B_{d22}\boldsymbol{\tau}_0}\right\}, \text{ and}$$
$$s_{\boldsymbol{\tau}_0,d} = \text{tr}(B_d) - (\boldsymbol{\tau}'_0\boldsymbol{\tau}_0)^{-1}\boldsymbol{\tau}'_0 B_d\boldsymbol{\tau}_0.$$

In addition, they also showed that equality holds in (5.3.9) if d is such that all B_{duv}, $1 \le u \le v \le 2$, are completely symmetric and d satisfies (5.3.6) or equivalently, (5.3.8). It is easy to see that for such a design d, $l_{\boldsymbol{\tau}_0,d}$ and $s_{\boldsymbol{\tau}_0,d}$ do not depend on $\boldsymbol{\tau}_0$. Now taking the average of $\phi_A(C_d, \boldsymbol{\tau}_0)$ over $\mathcal{P}$, they obtained a lower bound for the $\bar{\phi}_A$-criterion as shown in the next lemma.

Lemma 5.3.1. *For* $1 \le u \le v \le 2$, *let* $b_{duv} = \text{tr}(B_{duv})$. *Then*

$$\bar{\phi}_A(C_d) \ge \frac{(t-1)b_{d22}}{b_{d11}b_{d22} - b_{d12}^2} + \frac{(t-1)(t-2)}{b_{d11} + 2\gamma_0 b_{d12} + \gamma_0^2 b_{d22}}$$

and equality holds if d *is uniform on periods and all* B_{duv} *are completely symmetric,* $1 \le u \le v \le 2$. □

Recalling from Chapter 3 that a design is said to be binary over subjects if each treatment appears at most once in each subject, let $\mathcal{B}_{t,n,p}$ be the class of such designs in $\Omega_{t,n,p}$. Bailey and Kunert (2006) obtained the following result from Lemma 5.3.1.

Theorem 5.3.3. *Let* $d^* \in \Omega_{t,n,p}$, $p \le t$, *be a totally balanced design, and let the distribution* $\mathcal{P}$ *of* $\boldsymbol{\tau}_0$ *be permutation invariant.*
(i) Then d^* *is* $\bar{\phi}_A$*-optimal for direct effects over* $\mathcal{B}_{t,n,p}$ *for arbitrary* γ_0.
(ii) Suppose in addition, $\boldsymbol{\tau}_0$ *and* γ_0 *are independent.*
Then d^* *is* IA*-optimal for direct effects over* $\mathcal{B}_{t,n,p}$, *whatever be the distribution of* γ_0. □

Bailey and Kunert (2006) also considered a version of Lemma 5.3.1 in terms of the approximate theory and hence, using arguments somewhat similar to but more complex that those leading to Theorem 5.2.5, obtained two more optimality results on totally balanced designs. These are presented below.

Theorem 5.3.4. *Let* $d^* \in \Omega_{t,n,p}$ *be a totally balanced design, where* $t \ge p \ge 3$ *or* $p = 2, t \ge 3$. *Then for all* $\gamma_0 \in [-1, 1]$, d^* *is* $\bar{A}$*-optimal for direct effects over all designs in* $\Omega_{t,n,p}$ *which do not assign the same treatment to successive periods in any subject.* □

Theorem 5.3.5. *Let $d^* \in \Omega_{t,n,p}$ be a totally balanced design where $t \geq p \geq 3$ or $p = 2$ and $t \geq 6$. Then d^* is $\bar{A}$-optimal for direct effects over the entire class $\Omega_{t,n,p}$ if $\gamma_0 \in [-1, \varphi]$, where $\varphi(> 0)$ is given by*

$$\varphi = \frac{1}{(t-2)(p-1)}\left[(t-2) - \frac{1}{pt-t-1}\left(1 - \frac{t}{p(pt-t-1)}\right)^{-2}\right].$$

□

Bailey and Kunert (2006) also showed numerically that total balance does not imply $\bar{A}$-optimality for all values of $\boldsymbol{\tau}_0$ and γ_0, even over the binary designs. Moreover, for $p = 2, t = 5$ (a case not covered by Theorem 5.3.5), a totally balanced design is $\bar{A}$-optimal if $\gamma_0 \in [-0.45, 0.85]$. As γ_0 becomes large and positive, designs with $z_{dss} > 0$ (i.e., designs where a treatment follows itself) become better than totally balanced designs. Similar observations were made by Kempton *et al.* (2001) who used a computer algorithm to search for $\bar{A}$-optimal designs under model (5.3.2) in $\Omega_{4,12,4}$ and $\Omega_{4,12,5}$. They showed how the choice of the most efficient design for specified values of γ_0 changes with the value of γ_0 and designs with $z_{dss} = 0$ generally perform better when $\gamma_0 < 0$ while those with $z_{dss} > 0$ perform better for $\gamma_0 > 0$.

In the rest of this section, we consider optimal designs under (5.3.1) when γ is known. In this simpler case, the nonlinear model (5.3.1) becomes linear and the more stringent universal optimality criterion can be used for finding optimal designs for a given value of γ as considered by Bose and Stufken (2007). The linearity allows the development of analytical solutions for universally optimal designs for direct effects for certain broad ranges of possible values of γ.

For ease in presentation and in order to keep parity with the original paper, we now order the observations in $\boldsymbol{Y}_d$ as $\boldsymbol{Y}_d = (Y_{11}, Y_{12}, \ldots, Y_{np})'$. With this ordering of the observations in $\boldsymbol{Y}_d$, it follows that now the design matrices for period and subject effects are $P = I_p \otimes \mathbf{1}_n$ and $U = \mathbf{1}_p \otimes I_n$, respectively. Similarly, this ordering results in an appropriate rearrangement of the rows of the matrices T_d, F_d and W_d used earlier to give the corresponding matrices used in the rest of this section. Under model (5.3.1), for a design $d \in \Omega_{t,n,p}$ and a given value of γ, the information matrix for $\boldsymbol{\tau}$, which is denoted by $C_d(\gamma)$, is given by

$$C_d(\gamma) = (W_d(\gamma))'\mathrm{pr}^{\perp}([P\ U])W_d(\gamma). \qquad (5.3.10)$$

Definition 5.3.1. A design $d \in \Omega_{t,n,p}$ is said to be invariant over periods if for every treatment, the treatment is equally often replicated in each period; different treatments, however, may not have the same replication.

Clearly, a design that is uniform on periods is also invariant over the periods. Instead of studying $C_d(\gamma)$ directly, as in the previous section, it is again easier to study a simpler matrix that majorizes $C_d(\gamma)$ in the Loewner sense. To that end, we have the following lemma due to Bose and Stufken (2007). Note that in this lemma, equality holds under a condition weaker than that of uniformity on periods.

Lemma 5.3.2. *For a design $d \in \Omega_{t,n,p}$,*

$$C_d(\gamma) \le (W_d(\gamma))'\text{pr}^{\perp}([F_d\mathbf{1}_t \;\; U])W_d(\gamma)$$

in the Loewner sense, with equality if and only if d is invariant over the periods.

Proof. Since $F_d\mathbf{1}_t$ belongs to the column space of P, we have

$$\begin{aligned} C_d(\gamma) &= (W_d(\gamma))'\text{pr}^{\perp}([F_d\mathbf{1}_t \;\; U])W_d(\gamma) \\ &\quad -(W_d(\gamma))'\text{pr}(\text{pr}^{\perp}([F_d\mathbf{1}_t \;\; U])\text{pr}^{\perp}(F_d\mathbf{1}_t)P)W_d(\gamma) \\ &\le (W_d(\gamma))'\text{pr}^{\perp}([F_d\mathbf{1}_t \;\; U])W_d(\gamma), \end{aligned}$$

since $(W_d(\gamma))'\text{pr}(\text{pr}^{\perp}([F_d\mathbf{1}_t \;\; U])\text{pr}^{\perp}(F_d\mathbf{1}_t)P)W_d(\gamma)$ is nonnegative. Moreover, this term vanishes if and only if

$$(W_d(\gamma))'\text{pr}^{\perp}([F_d\mathbf{1}_t \;\; U])\text{pr}^{\perp}(F_d\mathbf{1}_t)P = \mathbf{0}. \tag{5.3.11}$$

It can be easily shown that

$$\text{pr}^{\perp}([F_d\mathbf{1}_t \;\; U])\text{pr}^{\perp}(F_d\mathbf{1}_t)P = \begin{bmatrix} \mathbf{0}_n & \mathbf{0}_n \cdots \mathbf{0}_n \\ \mathbf{0}_{n(p-1)} & \text{pr}^{\perp}(\mathbf{1}_{p-1}) \otimes \mathbf{1}_n \end{bmatrix}.$$

Hence (5.3.11) holds for all γ if and only if

$$T_d' \begin{bmatrix} \mathbf{0}_n & \mathbf{0}_n \cdots \mathbf{0}_n \\ \mathbf{0}_{n(p-1)} & \text{pr}^{\perp}(\mathbf{1}_{p-1}) \otimes \mathbf{1}_n \end{bmatrix} = F_d' \begin{bmatrix} \mathbf{0}_n & \mathbf{0}_n \cdots \mathbf{0}_n \\ \mathbf{0}_{n(p-1)} & \text{pr}^{\perp}(\mathbf{1}_{p-1}) \otimes \mathbf{1}_n \end{bmatrix} = \mathbf{0},$$

and this holds if and only if

$$\tilde{T}_{d1}'\mathbf{1}_n = \tilde{T}_{d2}'\mathbf{1}_n = \cdots = \tilde{T}_{dp}'\mathbf{1}_n, \tag{5.3.12}$$

where for $1 \le i \le p$, $\tilde{T}_{di}$ stands for the $n \times t$ submatrix of T_d corresponding to the observations in period i. Condition (5.3.12) is equivalent to d being invariant over the periods, thus completing the proof. □

The next step towards finding a universally optimal design is maximizing the trace of the upper bound of $C_d(\gamma)$ as given by Lemma 5.3.2. Writing

$$\begin{aligned} C_{d11} &= T_d'\mathrm{pr}^{\perp}(U)T_d, \\ C_{d12} &= T_d'\mathrm{pr}^{\perp}(U)F_d, \\ C_{d22} &= F_d'\mathrm{pr}^{\perp}(U)F_d, \end{aligned} \tag{5.3.13}$$

and $H_t = I_t - t^{-1}J_t$, we have the following result.

Lemma 5.3.3. *For any γ and for any design $d \in \Omega_{t,n,p}$,*

$$\mathrm{tr}(C_d(\gamma)) \le \mathrm{tr}(C_{d11}) + 2\gamma\mathrm{tr}(C_{d12}) + \gamma^2\mathrm{tr}(H_tC_{d22}), \tag{5.3.14}$$

with equality for all γ if d is uniform over periods.

Proof. Since $C_d(\gamma)\mathbf{1}_t = \mathbf{0}$, $\mathrm{tr}(C_d(\gamma)) = \mathrm{tr}(H_tC_d(\gamma))$. By virtue of Lemma 5.3.2, this implies that

$$\mathrm{tr}(C_d(\gamma)) = \mathrm{tr}(H_tC_d(\gamma)) \le \mathrm{tr}(H_t\left(W_d(\gamma))'\mathrm{pr}^{\perp}(F_d\mathbf{1}_t\ \ U\right)W_d(\gamma)). \tag{5.3.15}$$

By Lemma 1.2.1,

$$\begin{aligned}(W_d(\gamma))'\mathrm{pr}^{\perp}(F_d\mathbf{1}_t\ \ U)W_d(\gamma) = {} & (W_d(\gamma))'\mathrm{pr}^{\perp}(U)W_d(\gamma) \\ & -(W_d(\gamma))'\mathrm{pr}(\mathrm{pr}^{\perp}(U)F_d\mathbf{1}_t)W_d(\gamma).\end{aligned} \tag{5.3.16}$$

From (5.3.13) now observe that

$$(W_d(\gamma))'\mathrm{pr}^{\perp}(U)W_d(\gamma) = C_{d11} + \gamma(C_{d12} + C_{d12}') + \gamma^2C_{d22},$$

and thus

$$\mathrm{tr}\left(H_t(W_d(\gamma))'\mathrm{pr}^{\perp}(U)W_d(\gamma)\right) = \mathrm{tr}(C_{d11}) + 2\gamma\mathrm{tr}(C_{d12}) + \gamma^2\mathrm{tr}(H_tC_{d22}). \tag{5.3.17}$$

Since $\mathrm{tr}\left(H_t(W_d(\gamma))'\mathrm{pr}(\mathrm{pr}^{\perp}(U)F_d\mathbf{1}_t)W_d(\gamma)\right)$ is nonnegative, (5.3.14) follows from (5.3.15)–(5.3.17).

Equality holds in (5.3.14) if equality is attained in (5.3.15) and

$$\mathrm{tr}\left(H_t(W_d(\gamma))'\mathrm{pr}(\mathrm{pr}^{\perp}(U)F_d\mathbf{1}_t)W_d(\gamma)\right) = 0.$$

The last equation holds if and only if

$$H_t(W_d(\gamma))'\mathrm{pr}^{\perp}(U)F_d\mathbf{1}_t = \mathbf{0},$$

i.e., if and only if $(W_d(\gamma))'\mathrm{pr}^{\perp}(U)F_d\mathbf{1}_t$ is a multiple of $\mathbf{1}_t$.

Recalling that $W_d(\gamma) = T_d + \gamma F_d$, this happens for all γ if and only if each of $T_d'\mathrm{pr}^{\perp}(U)F_d\mathbf{1}_t$ and $F_d'\mathrm{pr}^{\perp}(U)F_d\mathbf{1}_t$ is a multiple of $\mathbf{1}_t$. Clearly, a design which is uniform on periods meets these requirements.

Moreover, such a design leads to equality in (5.3.15) for all γ. The result now follows. □

Remark 5.3.2. The only designs for which equality is attained in both Lemma 5.3.2 and Lemma 5.3.3 are the ones that are uniform on periods. □

The bound in Lemma 5.3.3 is important as it enables one to proceed as in the previous section to find an attainable upper bound for $\text{tr}(C_d(\gamma))$ which does not depend on d. For a design $d \in \Omega_{t,n,p}$, let T_{dj} and F_{dj} be the submatrices of T_d and F_d corresponding to subject j, and write

$$\begin{aligned} C_{d11}^{(j)} &= T'_{dj}\text{pr}^{\perp}(U)T_{dj}, \\ C_{d12}^{(j)} &= T'_{dj}\text{pr}^{\perp}(U)F_{dj}, \\ C_{d22}^{(j)} &= F'_{dj}\text{pr}^{\perp}(U)F_{dj}. \end{aligned}$$

Then it is easy to see as before that $C_{duv} = \sum_{j=1}^{n} C_{duv}^{(j)}$, for all u, v and

$$\begin{aligned} \text{tr}(C_{d11}) &= \sum_{j=1}^{n} \text{tr}(C_{d11}^{(j)}), \\ \text{tr}(C_{d12}) &= \sum_{j=1}^{n} \text{tr}(C_{d12}^{(j)}), \\ \text{tr}(H_t C_{d22}) &= \sum_{j=1}^{n} \text{tr}(H_t C_{d22}^{(j)}). \end{aligned} \tag{5.3.18}$$

Keeping in mind the expression in the right-hand side of (5.3.14), we now define the function

$$b_{\boldsymbol{s}}(\gamma) = \text{tr}(C_{11}^{\boldsymbol{s}}) + 2\gamma\text{tr}(C_{12}^{\boldsymbol{s}}) + \gamma^2\text{tr}(H_t C_{22}^{\boldsymbol{s}}), \tag{5.3.19}$$

where $C_{uv}^{\boldsymbol{s}}$ is used to denote $C_{duv}^{(j)}$ when the subject j receives the sequence $\boldsymbol{s} \in \mathcal{S}$. Clearly, in the expressions in (5.3.18), the contributions of subjects which receive sequences from the same equivalence class is the same. Hence $b_{\boldsymbol{s}}(\gamma)$ will be the same for all sequences $\boldsymbol{s}$ in the same equivalence class $< \boldsymbol{s} >$. So, from (5.3.18) it follows that for a given γ, in order to maximize the right-hand side of (5.3.14), i.e., $\text{tr}(C_{d11}) + 2\gamma\text{tr}(C_{d12}) + \gamma^2\text{tr}(H_t C_{d22})$, it is enough to choose sequences $\boldsymbol{s}$ which maximize $b_{\boldsymbol{s}}(\gamma)$ and then use designs d which allocate only those sequences which are equivalent to the chosen sequences.

Definition 5.3.2. For a given γ, a sequence will be called admissible for this value of γ if it maximizes $b_{\boldsymbol{s}}(\gamma)$ over all possible sequences $\boldsymbol{s}$ in $\mathcal{S}$. An equivalence class of admissible sequences will be called admissible.

Definition 5.3.3. A design $d \in \Omega_{t,n,p}$ is said to be neighbor balanced, if given any pair of distinct treatments s and s', d allocates them in consecutive periods to the same subject, irrespective of the order, an equal number of times; i.e., $z_{dss'} + z_{ds's}$ is a constant for all s, s', $1 \leq s, s' \leq t$, $s \neq s'$. No restrictions are placed on any z_{dss}.

Bose and Stufken (2007) proved the following result.

Theorem 5.3.6. *Assume a given value of γ in the model (5.3.1). Let $d^* \in \Omega_{t,n,p}$ be such that*

(i) d^* is uniform on periods;

(ii) all sequences in d^* are admissible for this value of γ, and

(iii) all the matrices $C_{d^*11}, (C_{d^*12} + C'_{d^*12})$ and C_{d^*22} are completely symmetric.

Then under model (5.3.1) with this given value of γ, d^* is universally optimal for direct effects over $\Omega_{t,n,p}$.

Proof. For any design $d \in \Omega_{t,n,p}$, Lemma 5.3.3, (5.3.18), (5.3.19) and conditions (i) and (ii) of the theorem together imply that

$$\begin{aligned} \operatorname{tr}(C_{d^*}(\gamma)) &= \operatorname{tr}(C_{d^*11} + 2\gamma C_{d^*12} + \gamma^2 H_t C_{d^*22}) \\ &= n \max_s b_{\mathbf{s}}(\gamma) \\ &\geq \operatorname{tr}(C_{d11} + 2\gamma C_{d12} + \gamma^2 H_t C_{d22}) \\ &\geq \operatorname{tr}(C_d(\gamma)). \end{aligned} \tag{5.3.20}$$

Thus the information matrix of d^* has maximal trace over all designs in $\Omega_{t,n,p}$. Also, from Lemma 5.3.2, (5.3.16) and condition (i) of the theorem, we have

$$\begin{aligned} C_{d^*}(\gamma) &= (W_{d^*}(\gamma))' \mathrm{pr}^{\perp}([F_{d^*}\mathbf{1}_t \;\; U]) W_{d^*}(\gamma) \\ &= (W_{d^*}(\gamma))' \mathrm{pr}^{\perp}(U) W_{d^*}(\gamma) - (W_{d^*}(\gamma))' \mathrm{pr}(\mathrm{pr}^{\perp}(U) F_{d^*}\mathbf{1}_t) W_{d^*}(\gamma) \\ &= C_{d^*11} + \gamma(C_{d^*12} + C'_{d^*12}) + \gamma^2 C_{d^*22} - c\mathbf{1}_t\mathbf{1}'_t, \end{aligned} \tag{5.3.21}$$

where c is a constant. Hence by condition (iii) of the theorem, $C_{d^*}(\gamma)$ is completely symmetric. The result is proved by invoking Theorem 1.4.1. □

Remark 5.3.3. Bose and Stufken (2007) pointed out that the conditions (i)–(iii) of Theorem 5.3.6 are sufficient for universal optimality but not necessary. For instance, if $\gamma = 0$, then in (iii), only the complete symmetry of C_{d^*11} is required. Again, for $t = 2$, condition (iii) is trivially true. It is easier to ensure that C_{d11} is completely symmetric if d is viewed as a block design with subjects as blocks and one keeps in mind that under the usual additive model for block designs, C_{d11} is the information matrix for treatment effects. The same is true for C_{d22} too, but now d needs to be restricted to the first $(p-1)$ periods and then viewed as a block design. Moreover, if d satisfies the other conditions, then $C_{d12}+C'_{d12}$ is completely symmetric if and only if d is neighbor balanced. So, the construction of designs satisfying conditions of Theorem 5.3.6 is not a difficult one. □

Bose and Stufken (2007) obtained the admissible sequences for some selected values of t and p with p in the practically useful range $2 \leq p \leq 4$ and used them to construct optimal designs satisfying the conditions of Theorem 5.3.6. Designs for other values of t and p may be similarly constructed. We present below three of the cases considered by Bose and Stufken (2007). In each case we give the equivalence classes, a typical sequence in each class and the corresponding value of $b_{\boldsymbol{S}}(\gamma)$. Then optimal designs are constructed using sequences from the admissible classes such that conditions (i) and (iii) hold. These examples are given for the smallest possible n for given t and p. Optimal designs with number of subjects a multiple of the n used in the examples can as usual be obtained by taking juxtapositions of the designs in the examples.

Case 1: $p=2,\ t \geq 2$.

In this case, there are only two equivalence classes.

Class	Sequence	$\mathrm{tr}(C_{11}^{\boldsymbol{S}})$	$\mathrm{tr}(C_{12}^{\boldsymbol{S}})$	$\mathrm{tr}(H_t C_{22}^{\boldsymbol{S}})$	$b_{\boldsymbol{S}}(\gamma)$
(1)	$(11)'$	0	0	0	0
(2)	$(12)'$	1	-0.5	0.25	$0.25(\gamma-2)^2$

Since $0.25(\gamma-2)^2 \geq 0$ for all γ, the equivalence class (2) is admissible for all γ. For $\gamma=2$, both the classes are admissible but $\tau_1-\tau_2$ is not estimable since $\max_{\boldsymbol{S}} b_{\boldsymbol{S}}(\gamma)=0$. So, for any other value of γ, sequences belonging to the class (2) should be used. If $t=2$, a design using the sequences $(12)'$ and $(21)'$ equally often is universally optimal for direct effects by virtue of Theorem 5.3.6. For odd $t \geq 3$, a neighbor balanced design that satisfies the conditions of Theorem 5.3.6 can be constructed when n is a multiple of $t(t-1)/2$. For even $t>2$, a similar result holds when n is a multiple of $t(t-1)$.

Example 5.3.1. With $t=3$ and $t=4$, the following designs d_1^* and d_2^*, based on sequences from class (2), satisfy conditions of Theorem 5.3.6. Hence d_1^* and d_2^* are universally optimal for direct effects over $\Omega_{3,3,2}$ and $\Omega_{4,12,2}$, respectively, for all values of γ.

$$d_1^* = \begin{matrix} 1\,2\,3 \\ 2\,3\,1 \end{matrix}, \quad d_2^* = \begin{matrix} 1\,1\,1\,2\,2\,2\,3\,3\,3\,4\,4\,4 \\ 2\,3\,4\,1\,3\,4\,1\,2\,4\,1\,2\,3 \end{matrix}.$$

□

Case 2: $p = 3, t = 2$.

In this case there are four equivalence classes.

Class	Sequence	$\mathrm{tr}(C_{11}^{\boldsymbol{S}})$	$\mathrm{tr}(C_{12}^{\boldsymbol{S}})$	$\mathrm{tr}(H_t C_{22}^{\boldsymbol{S}})$	$b_{\boldsymbol{S}}(\gamma)$
(1)	$(111)'$	0	0	0	0
(2)	$(112)'$	1.33	-0.33	0.33	$1.33 - 0.66\gamma + 0.33\gamma^2$
(3)	$(121)'$	1.33	-1	1	$1.33 - 2\gamma + \gamma^2$
(4)	$(122)'$	1.33	0	1	$1.33 + \gamma^2$

Here, if $\gamma > 0$, class (4) is the only admissible class and for $\gamma < 0$, class (3) is the only admissible class. If $\gamma = 0$, then classes (2), (3) and (4) are all admissible and any of these classes may be used for constructing optimal designs.

Example 5.3.2. Using sequences from class (4) and class (3), we construct designs d_1^* and d_2^*, respectively, shown below such that they satisfy conditions of Theorem 5.3.6. They are universally optimal for direct effects over $\Omega_{2,2,3}$ for any $\gamma \geq 0$ and $\gamma \leq 0$, respectively.

$$d_1^* = \begin{matrix} 1\,2 \\ 2\,1 \\ 2\,1 \end{matrix}, \quad d_2^* = \begin{matrix} 1\,2 \\ 2\,1 \\ 1\,2 \end{matrix}.$$

□

Case 3: $p = 3,\ t \geq 3$.

In this case, there are 5 equivalence classes.

Class	Sequence	$\mathrm{tr}(C_{11}^{\boldsymbol{S}})$	$\mathrm{tr}(C_{12}^{\boldsymbol{S}})$	$\mathrm{tr}(H_t C_{22}^{\boldsymbol{S}})$	$b_{\boldsymbol{S}}(\gamma)$
(1)	$(111)'$	0	0	0	0
(2)	$(112)'$	1.33	-0.33	0.33	$1.33 - 0.66\gamma + 0.33\gamma^2$
(3)	$(121)'$	1.33	-1	1	$1.33 - 2\gamma + \gamma^2$
(4)	$(122)'$	1.33	0	1	$1.33 + \gamma^2$
(5)	$(123)'$	2	-0.67	1.11	$2 - 1.34\gamma + 1.11\gamma^2$

It can be easily seen that (4) is the only admissible class when $\gamma \in (0.52, 11.48)$, (3) is the only admissible class when $\gamma \in (-4.73, -1.27)$ and (5) is admissible for all other values of γ.

Example 5.3.3. Let $t = 3$. Using sequences from class (4), class (3) and class (5), we construct designs d_1^*, d_2^*, and d_3^*, respectively, as shown below. These designs are universally optimal for direct effects over $\Omega_{3,3,3}$ for any γ in the ranges $(0.52, 11.48)$, $(-4.73, -1.27)$ and $(-1.27, 0.52)$, respectively.

$$d_1^* = \begin{matrix} 1 & 2 & 3 \\ 2 & 3 & 1 \\ 2 & 3 & 1 \end{matrix}, \quad d_2^* = \begin{matrix} 1 & 2 & 3 \\ 2 & 3 & 1 \\ 1 & 2 & 3 \end{matrix}, \quad d_3^* = \begin{matrix} 1 & 2 & 3 \\ 2 & 3 & 1 \\ 3 & 1 & 2 \end{matrix}.$$

□

Example 5.3.4. Let $t = 4$. The totally balanced design with 12 subjects given in Example 5.2.2 is universally optimal for direct effects over $\Omega_{4,12,3}$ for any γ in the range $(-1.27, 0.52)$. □

Remark 5.3.4. (i) As discussed in Remark 5.3.3, we try to ensure neighbor balance in the optimal designs and for this, we may not need to use all the sequences in an admissible class. For instance, in Example 5.3.3, we use only three sequences from class (5) instead of all the 6 sequences; in Example 5.3.4, 12 sequences out of the 24 in class (5) are used.
(ii) In the case $p = 6, t = 3$, it is known that the design d_2 in Example 2.2.2 is universally optimal under model (1.3.3) over $\Omega_{3,9,6}$ (see Theorem 2.4.1). However, under model (5.3.1), it can be seen that the optimal design with $n = 9$ will allocate the sequences $(112233)', (223311)'$, and $(331122)'$ to three subjects each. Thus the strongly balanced uniform design which has strong optimality properties under (1.3.3) may not always remain optimal under (5.3.1).
(iii) When $\gamma = 0$, there are no carryover effects and (5.3.1) reduces to the usual model for row-column designs. In the above examples, the optimal designs for $\gamma = 0$ match those for a row-column design, as expected. □

For other combinations of p and t, we can find optimal designs along similar lines. In Theorem 5.3.6, it was assumed that the constant of proportionality γ is known. In practice, γ will be *unknown* in most cases and will have to be estimated. This results in a loss of information for the estimation of treatment contrasts. It is thus important to examine whether or not the above findings obtained under the assumption that γ is known, are useful for the situation when γ is unknown. This issue was considered by Bose and Stufken (2007) and we present a summary of their results.

Recall that the information matrix for direct effects, C_d, under the model (5.3.2) is as given in (5.3.3). It can be shown from (5.3.3) and (5.3.10) that $C_d \leq C_d(\gamma_0)$ for all d, whatever be τ_0. Now, if d is uniform on periods, then, as in the proof of Lemma 5.3.2, it can be shown that

$$C_d = (W_d(\gamma_0))'\mathrm{pr}^{\perp}([F_d\mathbf{1}_t \quad U \quad F_d\boldsymbol{\tau}_0])W_d(\gamma_0). \tag{5.3.22}$$

Therefore, from Lemma 5.3.2 and (5.3.22), for any such design,

$$\begin{aligned} C_d(\gamma_0) - C_d &= (W_d(\gamma_0))'\{\mathrm{pr}^{\perp}([F_d\mathbf{1}_t \quad U]) - \mathrm{pr}^{\perp}([F_d\mathbf{1}_t \quad U \quad F_d\boldsymbol{\tau}_0])\}W_d(\gamma_0) \\ &= (W_d(\gamma_0))'\left\{\mathrm{pr}([F_d\mathbf{1}_t\ U \quad F_d\boldsymbol{\tau}_0]) - \mathrm{pr}([F_d\mathbf{1}_t \quad U])\right\}W_d(\gamma_0) \\ &= (W_d(\gamma_0))'\left\{\mathrm{pr}(\mathrm{pr}^{\perp}([F_d\mathbf{1}_t\ U])F_d\boldsymbol{\tau}_0)\right\}W_d(\gamma_0). \end{aligned}$$

Hence if one can identify a design d for which this difference is zero for some γ_0 and $\boldsymbol{\tau}_0$ and if d is universally optimal for direct effects when $\gamma = \gamma_0$, then d is also universally optimal if we treat γ as unknown and $\gamma_0, \boldsymbol{\tau}_0$ are the true values. In particular, we would like to find designs for which this difference is zero for some γ_0 whatever the value of $\boldsymbol{\tau}_0$. The following theorem of Bose and Stufken (2007) identifies such designs.

Theorem 5.3.7. *Let $d^* \in \Omega_{t,n,p}$ and suppose the true (unknown) value of γ is $\gamma_0 = -\mathrm{tr}(C_{d^*12})/\mathrm{tr}(H_tC_{d^*22})$. Furthermore, let d^* satisfy the following conditions:*

(i) d^* is uniform on periods,

(ii) all sequences in d^* are admissible for γ_0,

(iii) the matrices C_{d^*11}, C_{d^*12} and C_{d^*22} are all completely symmetric.
Then under model (5.3.1), d^* is universally optimal over $\Omega_{t,n,p}$ for the estimation of direct effects irrespective of the value of $\boldsymbol{\tau}_0$. □

For values of γ_0 other than that in Theorem 5.3.7, an optimal design may depend on the true value $\boldsymbol{\tau}_0$ of $\boldsymbol{\tau}$, making the problem of identifying universally optimal designs considerably harder. However, as remarked by Bose and Stufken (2007), the results for the case when γ is known can provide useful guidance for the case when γ is unknown.

Chapter 6

Optimality under Non-additive Models

6.1 Introduction

The optimal crossover designs considered in Chapters 2-4 are all obtained under a setup in which it is assumed that there is no interaction between treatments applied to the same subject in successive periods. While this assumption may be justified in some situations, the same may not be true in all cases where a crossover design is used. John and Quenouille (1977, pp. 211–214) gave an example of a crossover experiment with data on grass yields, and the analysis of these data show the presence of interaction among the treatments applied in successive periods on the same subject. Berenblut (1967) and Patterson (1970) considered crossover designs where the treatments were four equi-spaced levels of a single quantitative factor and the interaction between the linear component of direct effects and the linear component of carryover effects was important, apart from the (usual) direct and first order carryover effects. See also Berenblut (1968, 1970). Several designs for this situation were studied by Patterson (1970), including those proposed by Berenblut (1967). Patterson (1973), while describing a construction method of strongly balanced designs (see Section 2.6), remarked how these could be used in situations where the interaction between direct and carryover effects is included in the model, apart from the other parameters in model (1.2.1). Balaam (1968) also considered 2-period crossover designs under a model that included interaction between direct effects of treatments and periods. However, the question of optimality of these designs was not addressed by Balaam. See also a discussion by Federer of Hedayat (1981).

In view of the importance of interactions that might be present in certain situations, it is meaningful to study the optimality aspects of crossover designs in the presence of such an interaction. With this in mind, Sen and Mukerjee (1987) postulated a model by revising the additive model (1.2.1) to include an interaction term between the treatment producing the direct effect and the treatment producing the carryover effect in a given period of a subject. In this model, each treatment essentially has a different carryover effect for treatments (including itself) which follow it. Hedayat and Afsarinejad (2002) suggested a simpler model where each treatment has only two different carryover effects, one when it is followed by itself (self carryover) and another when it is followed by another treatment (mixed carryover). This self and mixed carryover model and associated optimal designs have already been discussed in Section 5.2. In this chapter we focus on results under the non-additive model as in Sen and Mukerjee (1987), where explicit interaction terms are incorporated.

John and Quenouille (1977) and Collombier and Merchermek (1993) used models where the carryover effects are not only due to the treatments applied to the immediately preceding period but also due to the treatments applied in earlier periods. Thus, in contrast to the model (1.2.1) where only the first order carryover effects are included, these authors considered the situation where carryover effects of orders higher than the first are possibly present. Bose and Mukherjee (2000, 2003) worked with a model incorporating carryover effects up to, say, k periods following the period of application. The presence of these higher order carryover effects also made it reasonable to allow the possible presence of the interactions between treatments in all these periods. Under such a model, Bose and Mukherjee (2000, 2003) obtained optimal crossover designs.

In obtaining optimality results under the above mentioned models, a calculus for factorial experiments is found to be a useful tool. In Section 6.2, we establish a correspondence between crossover designs and a symmetric t^2 factorial experiment, t being the number of treatments in the crossover design. For completeness and ready reference, we also state there some basic results from this calculus. Some optimality results for fixed subject effects under a non-additive model are given in Section 6.3 and the random subject effects case is studied in Section 6.4. Results on optimal crossover designs in the presence of higher order carryovers and interactions are summarized in Section 6.5. Relevant construction methods of optimal designs are given in the corresponding sections.

6.2 Correspondence with a Factorial Experiment

We incorporate an additional term in the model (1.2.1) to obtain the following non-additive model:

$$\begin{aligned} Y_{1j} &= \mu + \alpha_1 + \beta_j + \tau_{d(1,j)} + \epsilon_{1j},\ 1 \le j \le n, \\ Y_{ij} &= \mu + \alpha_i + \beta_j + \tau_{d(i,j)} + \rho_{d(i-1,j)} + \gamma_{d(i,j),d(i-1,j)} + \epsilon_{ij}, \\ &\qquad 2 \le i \le p,\ 1 \le j \le n, \end{aligned} \tag{6.2.1}$$

where $\gamma_{ss'}$ is the interaction effect between treatments s and s', $1 \le s, s' \le t$, and all other terms are as in (1.2.1).

Under the model (6.2.1), the response from a subject in any period other than the first period, is affected by the direct effect of a treatment, the first order carryover effect and their interaction, besides the period and subject effects. Under this model, it is very convenient to view a design $d \in \Omega_{t,n,p}$ as a two-factor symmetric t^2 factorial experiment arranged in a $p \times n$ row-column design, with the direct and first order carryover effects of treatments being looked upon, respectively, as the main effects of two factors F_1 and F_2, each at t levels, and the direct versus carryover effect interaction being given by the factorial two-factor interaction F_1F_2. Thus, instead of t treatments, we now need to consider the t^2 treatment combinations of the type (s, s'), $1 \le s, s' \le t$, where s, the treatment applied in the current period, represents the level of the first factor F_1 (contributing a direct effect) and s', the treatment applied in the immediately preceding period, represents the level of the second factor F_2 (contributing a carryover effect). A crossover design $d \in \Omega_{t,n,p}$ is therefore equivalent to a t^2 factorial experiment arranged in a row-column design with periods as rows and subjects as columns.

Let $\xi_{ss'}, 1 \le s, s' \le t$, denote the unknown parameter representing the effect of the treatment combination (s, s') described above. Write

$$\boldsymbol{\xi} = (\xi_{11}, \ldots, \xi_{1t}, \ldots, \xi_{t1}, \ldots, \xi_{tt})' \tag{6.2.2}$$

for the $t^2 \times 1$ vector of the effects of the t^2 factorial treatment combinations. Then model (6.2.1) may equivalently be written in terms of the $\xi_{ss'}$'s as

$$\begin{aligned} Y_{1j} &= \mu + \alpha_1 + \beta_j + t^{-1} \sum_{s'=1}^{t} \xi_{d(1,j)s'} + \epsilon_{1j},\quad 1 \le j \le n, \\ Y_{ij} &= \mu + \alpha_i + \beta_j + \xi_{d(i,j)d(i-1,j)} + \epsilon_{ij},\quad 2 \le i \le p,\ 1 \le j \le n, \end{aligned} \tag{6.2.3}$$

where the $\xi_{ss'}$, $1 \le s, s' \le t$, are as in (6.2.2) and other terms are as in (1.2.1). Note that in this non-circular model there are no carryover effects nor interactions contributing to the responses in the first period. In the factorial context this means that only the main effect of F_1 is involved in the responses Y_{1j}, leading to the averaging of the effects $\xi_{d(1,j),s'}$ over all the t levels of factor F_2.

The study of the properties of a design under model (6.2.3) becomes simpler if we use the tools of what is popularly known as the calculus for factorial arrangements. This calculus was first introduced by Kurkjian and Zelen (1962) and subsequently developed and studied by several others including, Kurkjian and Zelen (1963), Zelen and Federer (1964) and Mukerjee (1979, 1980). For a comprehensive discussion and results on the calculus for factorial arrangements with applications to design of factorial experiments, we refer to Gupta and Mukerjee (1989), where more references can be found; see also Dey and Mukerjee (1999). The following lemma, which is a two-factor version of a result in Mukerjee (1980), will be used later.

Lemma 6.2.1. *Consider a factorial arrangement with two factors, F_1 and F_2, each at t levels, in a design d. Let B_d be the coefficient matrix of the reduced normal equations for estimating the factorial treatment effects via the design d and let $Z^{10} = I_t \otimes J_t$ and $Z^{01} = J_t \otimes I_t$. Then in d,*
(i) the best linear unbiased estimators of contrasts belonging to the main effect F_1 are orthogonal to those belonging to the main effect F_2 and the interaction F_1F_2 if $Z^{10}B_d$ is symmetric, and
(ii) the best linear unbiased estimators of contrasts belonging to the main effect F_2 are orthogonal to those belonging to the main effect F_1 and the interaction F_1F_2 if $Z^{01}B_d$ is symmetric. □

We next develop the matrix B_d for a crossover design $d \in \Omega_{t,n,p}$ under the model (6.2.1) in terms of the equivalent factorial setup under the model (6.2.3). Let $\boldsymbol{e}_u$ be a $t\times 1$ vector with 1 in the uth position and zero elsewhere, $1 \le u \le t$. Then (6.2.3) may be written in the following equivalent form:

$$Y_{ij} = \mu + \alpha_i + \beta_j + \boldsymbol{\lambda}_{ij}'\boldsymbol{\xi} + \epsilon_{ij}, \quad 1 \le i \le p, \ 1 \le j \le n,$$

where

$$\begin{aligned} \boldsymbol{\lambda}_{1j} &= \boldsymbol{e}_{d(1,j)} \otimes t^{-1}\mathbf{1}_t, \\ \boldsymbol{\lambda}_{ij} &= \boldsymbol{e}_{d(i,j)} \otimes \boldsymbol{e}_{d(i-1,j)}, \quad 2 \le i \le p, \ 1 \le j \le n. \end{aligned} \tag{6.2.4}$$

Analogous to (1.3.3), the above model can also be written as

$$\mathbb{E}(\boldsymbol{Y}_d) = \tilde{X}_d\tilde{\boldsymbol{\theta}}, \qquad \mathbb{D}(\boldsymbol{Y}_d) = \sigma^2 I_{np}, \tag{6.2.5}$$

where $\boldsymbol{Y}_d$ is as in Section 1.3 and

$$\tilde{\boldsymbol{\theta}} = (\mu, \boldsymbol{\alpha}', \boldsymbol{\beta}', \boldsymbol{\xi}')'.$$

Furthermore, in (6.2.5), the design matrix $\tilde{X}_d$ is given by

$$\tilde{X}_d = [\mathbf{1}_{np} \;\; P \;\; U \;\; L],$$

where P and U are as in (1.3.4) and $L = (L_1', \ldots, L_n')'$, with $L_j = (\lambda_{1j}, \ldots, \lambda_{pj})'$, $1 \le j \le n$.

Example 6.2.1. Under the model (6.2.5), the design d_1 of Example 2.2.1 is equivalent to a 4^2 factorial experiment. Two typical $\boldsymbol{\lambda}_{ij}$ vectors are as follows:

$$\begin{aligned}
\boldsymbol{\lambda}_{13} &= (0,0,1,0)' \otimes \frac{1}{4}(1,1,1,1)' \\
&= \frac{1}{4}(0,0,0,0,0,0,0,0,1,1,1,1,0,0,0,0)' \\
\boldsymbol{\lambda}_{34} &= (1,0,0,0)' \otimes (0,0,1,0)' \\
&= (0,0,1,0,0,0,0,0,0,0,0,0,0,0,0,0)'.
\end{aligned}$$

□

From (6.2.5), similar to (1.3.7), we have for any $d \in \Omega_{t,n,p}$,

$$\tilde{X}_d'\tilde{X}_d = \begin{bmatrix} np & n\mathbf{1}_p' & p\mathbf{1}_n' & \sum_{i=1}^{p}\sum_{j=1}^{n} \boldsymbol{\lambda}_{ij}' \\ n\mathbf{1}_p & nI_p & J_{pn} & \tilde{M}_d' \\ p\mathbf{1}_n & J_{np} & pI_n & \tilde{N}_d' \\ \sum_{i=1}^{p}\sum_{j=1}^{n} \boldsymbol{\lambda}_{ij} & \tilde{M}_d & \tilde{N}_d & \tilde{V}_d \end{bmatrix}, \tag{6.2.6}$$

where

$$\begin{aligned}
\tilde{V}_d &= \sum_{i=1}^{p}\sum_{j=1}^{n} \boldsymbol{\lambda}_{ij}\boldsymbol{\lambda}_{ij}', \\
\tilde{M}_d &= \left(\sum_{j=1}^{n} \boldsymbol{\lambda}_{1j}, \ldots, \sum_{j=1}^{n} \boldsymbol{\lambda}_{pj}\right), \qquad (6.2.7) \\
\tilde{N}_d &= \left(\sum_{i=1}^{p} \boldsymbol{\lambda}_{i1}, \ldots, \sum_{i=1}^{p} \boldsymbol{\lambda}_{in}\right).
\end{aligned}$$

Thus $\tilde{M}_d$ and $\tilde{N}_d$ are respectively the "treatments" versus periods and "treatments" versus subjects incidence matrices of orders $t^2 \times p$ and $t^2 \times n$, respectively, where the t^2 treatment combinations of the form (s, s'), $1 \le s, s' \le t$, play the role of "treatments".

Now, as in Section 1.3, from (6.2.6) it follows that the coefficient matrix of the reduced normal equations for estimating $\boldsymbol{\xi}$ via a design $d \in \Omega_{t,n,p}$ is given by

$$B_d = \tilde{V}_d - n^{-1}\tilde{M}_d\tilde{M}_d' - p^{-1}\tilde{N}_d\tilde{N}_d' + (np)^{-1}(\tilde{N}_d\mathbf{1}_n)(\tilde{N}_d\mathbf{1}_n)'. \qquad (6.2.8)$$

However, it is to be noted that we are not interested in estimating all contrasts among the components of $\boldsymbol{\xi}$; rather, we are interested in estimating contrasts among direct and carryover effects or, equivalently, contrasts belonging to the main effects F_1 and F_2. Therefore it is necessary to define contrasts belonging to these main effects in terms of $\boldsymbol{\xi}$. It is easy to see that typical contrasts belonging to the main effect F_1, main effect F_2 and the interaction F_1F_2 are given by

$$(\boldsymbol{l}_1 \otimes \mathbf{1}_t)'\boldsymbol{\xi}, \quad (\mathbf{1}_t \otimes \boldsymbol{l}_2)'\boldsymbol{\xi} \text{ and } (\boldsymbol{l}_1 \otimes \boldsymbol{l}_2)'\boldsymbol{\xi},$$

respectively, where $\boldsymbol{l}_1$ and $\boldsymbol{l}_2$ are any $t \times 1$ nonnull vectors satisfying $\boldsymbol{l}_1'\mathbf{1}_t = 0 = \boldsymbol{l}_2'\mathbf{1}_t$.

6.3 Optimality Results

Theorem 2.4.1 shows that strongly balanced uniform designs are universally optimal under (1.2.1) for both direct and carryover effects. Sen and Mukerjee (1987) showed that these designs remain optimal for direct effects even under model (6.2.1). This is primarily because the direct effect contrasts are estimable orthogonally to those belonging to carryover effects and interactions.

To prove this result, one could obtain the information matrix for the estimation of direct effects from B_d and then show that this information matrix satisfies the sufficient conditions of universal optimality as given in Theorem 1.4.1. Alternatively, by noting that the model (6.2.1) is "finer" than the model (1.2.1) and then applying Lemma 1.4.1 and Theorem 2.4.1, one can get a simpler proof as in Sen and Mukerjee (1987). We present this proof here.

Theorem 6.3.1. *Let d^* be a strongly balanced uniform design in $\Omega_{t,n,p}$. Then under the model (6.2.1), d^* is universally optimal for the estimation of direct effects over $\Omega_{t,n,p}$.*

Proof. By Theorem 2.4.1, d^* is universally optimal for direct effects under the additive model (1.2.1). So, under model (6.2.1), which is the non-additive version of (1.2.1), in order to prove the optimality of d^* we need to show that appropriate orthogonality conditions hold, i.e., it suffices to show that in d^*, contrasts belonging to main effect F_1 are estimable orthogonally to those belonging to main effect F_2 and interaction F_1F_2. Now, by Lemma 6.2.1, this will be proved if one can show that $Z^{10}B_{d^*}$ is symmetric. To achieve this, we first evaluate $\tilde{V}_{d^*}, \tilde{M}_{d^*}$ and $\tilde{N}_{d^*}$ as given by (6.2.7) using the properties of d^* given in Definitions 2.2.3 and 2.2.5.

For d^*, we have from (6.2.4),

$$\begin{aligned}\sum_{j=1}^{n} \boldsymbol{\lambda}_{1j}\boldsymbol{\lambda}'_{1j} &= \sum_{j=1}^{n} \boldsymbol{e}_{d^*(1,j)}\boldsymbol{e}'_{d^*(1,j)} \otimes t^{-2}J_t \\ &= \frac{n}{t^3}\left(I_t \otimes J_t\right),\end{aligned}$$

as d^* is uniform over period 1, i.e., each treatment appears n/t times in period 1. Similarly,

$$\begin{aligned}\sum_{i=2}^{p}\sum_{j=1}^{n} \boldsymbol{\lambda}_{ij}\boldsymbol{\lambda}'_{ij} &= \sum_{i=2}^{p}\sum_{j=1}^{n} \boldsymbol{e}_{d^*(i,j)}\boldsymbol{e}'_{d^*(i,j)} \otimes \boldsymbol{e}_{d^*(i-1,j)}\boldsymbol{e}'_{d^*(i-1,j)} \\ &= \frac{n(p-1)}{t^2}\left(I_t \otimes I_t\right),\end{aligned}$$

since d^* is strongly balanced, i.e., each pair of treatments occurs $n(p-1)/t^2$ times in consecutive periods in the same subject over periods $2, \ldots, p$.

Thus from (6.2.7),

$$\tilde{V}_{d^*} = \frac{n}{t^3}(I_t \otimes J_t) + \frac{n(p-1)}{t^2}(I_t \otimes I_t). \tag{6.3.1}$$

Again, for d^*,

$$\sum_{j=1}^{n} \boldsymbol{\lambda}_{1j} = \frac{n}{t^2}(\mathbf{1}_t \otimes \mathbf{1}_t) \text{ and } \sum_{i=2}^{p}\sum_{j=1}^{n} \boldsymbol{\lambda}_{ij} = \frac{n(p-1)}{t^2}(\mathbf{1}_t \otimes \mathbf{1}_t), \tag{6.3.2}$$

because of uniformity of d^* over periods and strong balance. Thus

$$\tilde{N}_{d^*}\mathbf{1}_n = \sum_{i=1}^{p}\sum_{j=1}^{n} \boldsymbol{\lambda}_{ij} = \frac{np}{t^2}(\mathbf{1}_t \otimes \mathbf{1}_t). \tag{6.3.3}$$

Also, from (6.2.7),

$$Z^{10}\tilde{M}_{d^*} = (I_t \otimes J_t)\left[\sum_{j=1}^{n} \boldsymbol{e}_{d^*(1,j)} \otimes \frac{1}{t}\mathbf{1}_t, \sum_{j=1}^{n} \boldsymbol{e}_{d^*(2,j)} \otimes \boldsymbol{e}_{d^*(1,j)}, \right.$$

$$\left. \ldots, \sum_{j=1}^{n} \boldsymbol{e}_{d^*(p,j)} \otimes \boldsymbol{e}_{d^*(p-1,j)}\right]$$

$$= \left[\sum_{j=1}^{n} \boldsymbol{e}_{d^*(1,j)} \otimes \mathbf{1}_t, \sum_{j=1}^{n} \boldsymbol{e}_{d^*(2,j)} \otimes \mathbf{1}_t, \ldots, \sum_{j=1}^{n} \boldsymbol{e}_{d^*(p,j)} \otimes \mathbf{1}_t\right]$$

$$= \left[\frac{n}{t}(\mathbf{1}_t \otimes \mathbf{1}_t), \ldots, \frac{n}{t}(\mathbf{1}_t \otimes \mathbf{1}_t)\right],$$

as d^* is uniform over periods. Hence on simplification, using (6.3.2),

$$\begin{aligned} Z^{10}\tilde{M}_{d^*}\tilde{M}'_{d^*} &= \frac{n}{t}\left[(\mathbf{1}_t \otimes \mathbf{1}_t)\sum_{j=1}^{n} \boldsymbol{\lambda}'_{1j} + \sum_{i=2}^{p}(\mathbf{1}_t \otimes \mathbf{1}_t)\sum_{j=1}^{n} \boldsymbol{\lambda}'_{ij}\right] \\ &= \frac{n^2 p}{t^3}(J_t \otimes J_t). \end{aligned} \tag{6.3.4}$$

Similarly, from (6.2.4),

$$\begin{aligned} Z^{10}\sum_{i=1}^{p} \boldsymbol{\lambda}_{ij} &= (I_t \otimes \mathbf{1}_t)(I_t \otimes \mathbf{1}'_t)\left(\boldsymbol{e}_{d^*(1,j)} \otimes t^{-1}\mathbf{1}_t + \sum_{i=2}^{p} \boldsymbol{e}_{d^*(i,j)} \otimes \boldsymbol{e}_{d^*(i-1,j)}\right) \\ &= (I_t \otimes \mathbf{1}_t)\left(\boldsymbol{e}_{d^*(1,j)} + \sum_{i=2}^{p} \boldsymbol{e}_{d^*(i,j)}\right) \\ &= (I_t \otimes \mathbf{1}_t)\sum_{i=1}^{p} \boldsymbol{e}_{d^*(i,j)} = pt^{-1}(\mathbf{1}_t \otimes \mathbf{1}_t),\ 1 \leq j \leq n, \end{aligned}$$

since d^* is uniform over subjects. Hence from (6.2.7),

$$Z^{10}\tilde{N}_{d^*} = pt^{-1}(\mathbf{1}_t \otimes \mathbf{1}_t, \ldots, \mathbf{1}_t \otimes \mathbf{1}_t)$$

and so, using (6.3.3),

$$Z^{10}\tilde{N}_{d^*}\tilde{N}'_{d^*} = \frac{p}{t}(\mathbf{1}_{t^2}\mathbf{1}'_n)\tilde{N}'_{d^*} = \frac{np^2}{t^3}(J_t \otimes J_t). \tag{6.3.5}$$

From (6.2.8), (6.3.1), (6.3.3), (6.3.4) and (6.3.5), it is clear that $Z^{10}B_{d^*}$ is symmetric and the proof is complete. □

Theorem 6.3.1 does not have an exact counterpart for the estimation of carryover effects. This is because, unlike in the case of direct effects, not all strongly balanced uniform designs satisfy the appropriate orthogonality condition. The following example illustrates this point.

Example 6.3.1. Consider the two strongly balanced uniform designs d_1 and d_2, both belonging to $\Omega_{2,4,6}$, as shown below.

$$d_1 = \begin{matrix} 1 & 1 & 2 & 2 \\ 1 & 2 & 1 & 2 \\ 1 & 2 & 2 & 1 \\ 2 & 2 & 1 & 1 \\ 2 & 1 & 2 & 1 \\ 2 & 1 & 1 & 2 \end{matrix}, \quad d_2 = \begin{matrix} 2 & 1 & 1 & 2 \\ 1 & 1 & 2 & 2 \\ 1 & 1 & 2 & 2 \\ 2 & 2 & 1 & 1 \\ 1 & 2 & 2 & 1 \\ 2 & 2 & 1 & 1 \end{matrix}.$$

The matrices B_{d_1} and B_{d_2} can be computed from (6.2.8) and it can be easily verified that while $Z^{01}B_{d_2}$ is symmetric, $Z^{01}B_{d_1}$ is not so. By Lemma 6.2.1, this implies that d_2 allows estimation of carryover effects orthogonally to direct effects and direct versus carryover effects interaction; however, such a property does not hold for d_1, even though both d_1 and d_2 are strongly balanced uniform designs. □

In view of Example 6.3.1, one needs to identify the strongly balanced uniform designs which allow the estimation of carryover effect contrasts orthogonally to those belonging to direct effect and direct versus carryover interaction. A solution to this problem essentially requires a combinatorial characterization of the symmetry of $Z^{01}B_{d^*}$. A complete characterization of this kind is, however, technically too involved to be helpful in actual construction of designs. Instead, one may look for simpler sufficient conditions. A set of such sufficient conditions was obtained by K. L. Kok in an unpublished Ph.D. thesis (see Patterson (1973)) for the special case of $n = t^2, p = 2t$. A more general set of sufficient conditions was given by Sen and Mukerjee (1987) as shown in Lemma 6.3.1 below.

Lemma 6.3.1. *For a design $d \in \Omega_{t,n,p}$, let ζ_{ds} be the set of subjects receiving treatment s in the last period, $1 \le s \le t$. Let $d^* \in \Omega_{t,n,p}$ be a strongly balanced uniform design. Then under model (6.2.1), d^* allows orthogonal estimation of carryover effect contrasts if it satisfies the following conditions:*

(i) for each s, s' $(1 \le s, s' \le t)$, there are exactly n/t^2 subjects receiving treatments s and s' in the first and last periods, respectively, and

(ii) for each s $(1 \le s \le t)$, in the collection of ordered pairs $\{d^*(i-1,j), d^*(i,j)\}$, $2 \le i \le p$, $j \in \zeta_{d^*s}$, each ordered pair (s, s_2) $(1 \le s_2 \le t)$ occurs the same number (say, u_1) of times while each ordered pair

(s_1, s_2), $(1 \le s_1, s_2 \le t, s_1 \ne s)$ occurs the same number (say u_2) of times.

Proof. As in the proof of Theorem 6.3.1, for every strongly balanced uniform design d^*, it may be checked using (6.3.1)–(6.3.3) that each of the matrices

$$Z^{01}\tilde{V}_{d^*}, \quad Z^{01}\tilde{M}_{d^*}\tilde{M}'_{d^*} \quad \text{and} \quad Z^{01}(\tilde{N}_{d^*}\mathbf{1}_n)(\tilde{N}_{d^*}\mathbf{1}_n)'$$

is symmetric. Thus by Lemma 6.2.1 and (6.2.8), it suffices to show that $Z^{01}\tilde{N}_{d^*}\tilde{N}'_{d^*}$ is symmetric when d^* satisfies the conditions (i) and (ii) of the lemma. From (6.2.7), we have

$$\tilde{N}_{d^*}\tilde{N}'_{d^*} = \sum_{j=1}^{n}\left(\sum_{i=1}^{p}\boldsymbol{\lambda}_{ij}\right)\left(\sum_{i=1}^{p}\boldsymbol{\lambda}'_{ij}\right). \tag{6.3.6}$$

Also, for d^*, from (6.2.4),

$$Z^{01}\left(\sum_{i=1}^{p}\boldsymbol{\lambda}_{ij}\right) = \mathbf{1}_t \otimes \left[t^{-1}\mathbf{1}_t + \sum_{i=2}^{p}\mathbf{e}_{d^*(i-1,j)}\right]$$

$$= \mathbf{1}_t \otimes \left[(p+1)t^{-1}\mathbf{1}_t - \mathbf{e}_{d^*(p,j)}\right],$$

since $\sum_{i=1}^{p}\mathbf{e}_{d^*(i,j)} = \frac{p}{t}\mathbf{1}_t$. Hence from (6.3.6),

$$\begin{aligned} Z^{01}\tilde{N}_{d^*}\tilde{N}'_{d^*} = \frac{p+1}{t}(\mathbf{1}_t \otimes \mathbf{1}_t)\left(\sum_{j=1}^{n}\sum_{i=1}^{p}\boldsymbol{\lambda}'_{ij}\right) \\ - \sum_{j=1}^{n}\left(\mathbf{1}_t \otimes \mathbf{e}_{d^*(p,j)}\right)\left(\sum_{i=1}^{p}\boldsymbol{\lambda}'_{ij}\right). \end{aligned} \tag{6.3.7}$$

The first term on the right-hand side of (6.3.7) is symmetric by (6.3.2). The second term equals

$$\begin{aligned} & t^{-1}\sum_{j=1}^{n}\left(\mathbf{1}_t\mathbf{e}'_{d^*(1,j)}\right) \otimes \left(\mathbf{e}_{d^*(p,j)}\mathbf{1}'_t\right) \\ & \qquad + \sum_{j=1}^{n}\left(\mathbf{1}_t \otimes \mathbf{e}_{d^*(p,j)}\right)\left(\sum_{i=2}^{p}\mathbf{e}'_{d^*(i,j)} \otimes \mathbf{e}'_{d^*(i-1,j)}\right) \\ & = \frac{n}{t^3}(J_t \otimes J_t) + \sum_{s=1}^{t}(\mathbf{1}_t \otimes \mathbf{e}_s)\left(\sum_{j\in\zeta_{d^*s}}\sum_{i=2}^{p}\mathbf{e}'_{d^*(i,j)} \otimes \mathbf{e}'_{d^*(i-1,j)}\right) \\ & = \frac{n}{t^3}(J_t \otimes J_t) + \sum_{s=1}^{t}(\mathbf{1}_t \otimes \mathbf{e}_s)\left[\mathbf{1}'_t \otimes \{u_2\mathbf{1}'_t - (u_2 - u_1)\mathbf{e}'_s\}\right] \\ & = \left(\frac{n}{t^3} + u_2\right)(J_t \otimes J_t) - (u_2 - u_1)(J_t \otimes I_t), \end{aligned}$$

by virtue of the conditions (i) and (ii) for d^*. It is now easy to see that $Z^{01}B_{d^*}$ is symmetric and the result follows. □

From Theorem 2.4.1 and Lemma 6.3.1, by applying Lemma 1.4.1, the following result is immediate.

Theorem 6.3.2. *Let $d^* \in \Omega_{t,n,p}$ be a strongly balanced uniform design satisfying conditions (i) and (ii) of Lemma 6.3.1. Then d^* is universally optimal for carryover effects over $\Omega_{t,n,p}$.* □

Remark 6.3.1. If d^* satisfies condition (ii) of Lemma 6.3.1, then by Definitions 2.2.3 and 2.2.5 it follows that $u_1 = n(p-t)/t^3$ and $u_2 = np/t^3$. □

Remark 6.3.2. Suppose $\mu_2 = p/t$ is even. Then the method of construction in Theorem 2.6.2 can be applied with $\alpha = 0$ and $\beta = \mu_2/2$. It is not hard to see that the resulting designs satisfy the condition of Lemma 6.3.1 and are therefore universally optimal, by Theorem 6.3.2, for the carryover effects under the nonadditive model. The design d_2 in Example 2.2.1 is one such design. When μ_2 is odd and $t \neq 6$, the following result from Sen and Mukerjee (1987) provides a method of construction of designs satisfying both conditions of Lemma 6.3.1. □

Theorem 6.3.3. *For $t \neq 6$, a strongly balanced uniform design satisfying conditions (i) and (ii) of Lemma 6.3.1 exists whenever $n = \mu_3 t^2$ and $p = \mu_2 t$ for integers $\mu_3 \geq 1$ and $\mu_2 > 2$ odd.*

Proof. Let $\mu_2 = 2m+1$ where m is a positive integer. First suppose $t \neq 2, 6$. Then a pair of mutually orthogonal $t \times t$ Latin squares exists. Let these Latin squares be denoted by Q_1 and Q_2 and without loss of generality, let the symbols in these squares be $0, 1, \ldots, t-1$. Let $\boldsymbol{q}_{uh}$ be the hth column of $Q_u (u = 1, 2)$, $\boldsymbol{g}_h$ be the $t \times 1$ vector with all elements equal to h and G_h be the $t \times 3$ matrix given by $G_h = (\boldsymbol{q}_{1h}, \boldsymbol{q}_{2h}, \boldsymbol{g}_h)$, $0 \leq h \leq t-1$. Define

$$G = (G_0, G_1, \cdots, G_{t-1}).$$

For $t = 2$, define G as

$$G = \begin{bmatrix} 1\,0\,0\,1\,0\,1 \\ 0\,0\,0\,1\,1\,1 \end{bmatrix}.$$

For any $t \neq 6$, now write

$$B_0 = \begin{bmatrix} 0\;1 \cdots\; t-1 \\ 0\;0 \cdots \quad 0 \end{bmatrix}',$$

and for $1 \leq h \leq t-1$, let B_h be obtained by adding h to each element of B_0, where the elements are reduced mod t. Let $B = [B_0, B_1, \ldots, B_{t-1}]$ and define the $t \times p$ array

$$A_0 = [G, \underbrace{B, \ldots, B}_{m-1 \text{ times}}].$$

For $1 \leq h \leq t-1$, let A_h be obtained by adding h to each element of A_0, where the elements are reduced mod t. Then it can be verified that the $p \times t^2$ array

$$A = [A'_0, A'_1, \ldots, A'_{t-1}]$$

with rows and columns identified with periods and subjects, respectively, is a strongly balanced uniform crossover design d^* in $\Omega_{t,t^2,p}$ that satisfies the conditions (i) and (ii) of Lemma 6.3.1. A strongly balanced uniform design in $\Omega_{t,\mu_3 t^2,p}$ satisfying the same conditions can now be obtained by taking μ_3 copies of the design d^* constructed above. □

Example 6.3.2. Let $t = 2$. If the method of construction in Theorem 6.3.3 is applied with $m = 1$ and in the resulting array A, the entries 0 and 1 are replaced by 1 and 2, respectively, then we get the design d_2 in Example 6.3.1. □

Example 6.3.3. Let $t = 3$ and $p = 15$, i.e., $m = 2$. For this case, we illustrate in detail the construction of d^* by the above method. Consider the two orthogonal Latin squares of order 3 as

$$Q_1 = \begin{pmatrix} 0\,1\,2 \\ 1\,2\,0 \\ 2\,0\,1 \end{pmatrix}, \quad Q_2 = \begin{pmatrix} 0\,1\,2 \\ 2\,0\,1 \\ 1\,2\,0 \end{pmatrix}.$$

Following the construction given in the proof of Theorem 6.3.3, we have

$$B = \begin{bmatrix} 0\,0 & 1\,1 & 2\,2 \\ 1\,0 & 2\,1 & 0\,2 \\ 2\,0 & 0\,1 & 1\,2 \end{bmatrix}, \quad G = \begin{bmatrix} 0\,0\,0 & 1\,1\,1 & 2\,2\,2 \\ 1\,2\,0 & 2\,0\,1 & 0\,1\,2 \\ 2\,1\,0 & 0\,2\,1 & 1\,0\,2 \end{bmatrix}.$$

Then $A_0 = (G, B)$ and A_1 and A_2 are obtained by adding 1 and 2 respectively to each element of A_0 and reducing the elements mod 3. The 15×9 array $A = (A'_0, A'_1, A'_2)$ gives the required strongly balanced uniform design in $\Omega_{3,9,15}$ that satisfies the conditions of Lemma 6.3.1. □

Remark 6.3.3. Sen and Mukerjee (1987) also showed by actual construction that if one ignores the conditions of Lemma 6.3.1, then a strongly balanced uniform crossover design exists also when $t = 6$, p is an odd multiple of 6 and $t^2|n$. □

In a sense, Theorems 6.3.1 and 6.3.2 establish the robustness of Theorem 2.4.1 under a non-additive model. The next result revisits Theorem 2.4.2 from this perspective.

Theorem 6.3.4. *Let $d^* \in \Omega_{t,n,p}$ be a strongly balanced design that is uniform on the periods and uniform on subjects in the first $(p-1)$ periods. Then under model (6.2.3), d^* is universally optimal for carryover effects over $\Omega_{t,n,p}$.*

Proof. In view of Theorem 2.4.2, it suffices to check the relevant orthogonality condition as given in Lemma 6.2.1, namely, the symmetry of $Z^{01}B_{d^*}$. From the definition of d^*, this symmetry can be verified along the lines of the proofs of Theorems 6.3.1 and 6.3.2. □

Remark 6.3.4. Theorem 6.3.4 shows that if the carryover effects are of primary interest, one may use smaller sized designs as the condition of uniformity over subjects is no longer required. □

6.4 Optimality Under a Non-additive Random Subject Effects Model

Some authors, e.g., Laska and Meisner (1985), Carriere and Reinsel (1993) and Hedayat *et al.* (2006) studied the situations where the subject effects are random. They obtained optimal designs under a version of model (1.2.1) assuming random subject effects (see Sections 3.2, 3.3 and 7.2). In this section we study a version of model (6.2.1) under such an assumption. So, with Y_{ij}'s as in model (6.2.1), we assume that the vector of subject effects $\boldsymbol{\beta} = (\beta_1, \ldots, \beta_n)'$ has a distribution with mean $\mathbf{0}$ and dispersion matrix $\sigma_\beta^2 I_n$ while the vector of error terms has a distribution with mean $\mathbf{0}$ and dispersion matrix $\sigma^2 I_{np}$. It is further assumed that $\boldsymbol{\beta}$ is independent of the vector of error components.

Note that the above described linear model with random subject effects may be written analogously to (6.2.5) as

$$\mathbb{E}(\boldsymbol{Y}_d) = \tilde{X}_d\tilde{\boldsymbol{\theta}}, \qquad \mathbb{D}(\boldsymbol{Y}_d) = \sigma^2\Sigma, \tag{6.4.1}$$

where $\tilde{X}_d$ and $\tilde{\boldsymbol{\theta}}$ are now defined as

$$\tilde{X}_d = [\mathbf{1}_{np} \;\; P \;\; L], \qquad \tilde{\boldsymbol{\theta}} = (\mu, \;\; \boldsymbol{\alpha}', \;\; \boldsymbol{\xi}')',$$

and $\Sigma = I_n \otimes V$ with

$$V = I_p + \frac{\sigma_\beta^2}{\sigma^2} J_p.$$

Considering weighted least squares estimation based on $\tilde{X}_d' \Sigma^{-1} \tilde{X}_d$, after some algebra one can obtain the coefficient matrix of the reduced normal equations for estimating $\boldsymbol{\xi}$ via a design $d \in \Omega_{t,n,p}$ under this model. Let this matrix be denoted by $\tilde{B}_d$. The following lemma shows $\tilde{B}_d$ explicitly.

Lemma 6.4.1. *For any design* $d \in \Omega_{t,n,p}$,

$$\begin{aligned} \tilde{B}_d &= \frac{1}{\sigma^2} \left[\tilde{V}_d - \frac{1}{n} \tilde{M}_d \tilde{M}_d' - \frac{\sigma_\beta^2}{\sigma^2 + p\sigma_\beta^2} \tilde{N}_d H_n \tilde{N}_d' \right] \\ &= \frac{1}{\sigma^2} B_d + \frac{1}{p(\sigma^2 + p\sigma_\beta^2)} \tilde{N}_d H_n \tilde{N}_d', \end{aligned}$$

where $\tilde{V}_d$, $\tilde{M}_d$, $\tilde{N}_d$ *are as in (6.2.7),* B_d *is as in (6.2.8) and* $H_n = I_n - n^{-1} J_n$. □

Now, in order to express the contrasts belonging to direct effects, carryover effects and their interactions using matrix notation, we first consider a $(t-1) \times t$ matrix, P_t, such that $(t^{-\frac{1}{2}} \mathbf{1}_t, P_t')$ is an orthogonal matrix. Then define

$$P^{01} = (t^{-\frac{1}{2}} \mathbf{1}_t') \otimes P_t, \; P^{10} = P_t \otimes (t^{-\frac{1}{2}} \mathbf{1}_t'), \; P^{11} = P_t \otimes P_t. \tag{6.4.2}$$

Note that with $\boldsymbol{\xi}$ as in (6.2.2), $P^{01}\boldsymbol{\xi}$, $P^{10}\boldsymbol{\xi}$ and $P^{11}\boldsymbol{\xi}$ represent complete sets of orthonormal contrasts belonging to the carryover effects, the direct effects and their interaction, respectively. Hence following Mukerjee (1980), it can be shown that for a design $d \in \Omega_{t,n,p}$, the coefficient matrix of the reduced normal equations for estimating the carryover effects is given by

$$A_d = P^{01} \tilde{B}_d (P^{01})' - [P^{01} \tilde{B}_d (P^{10})', P^{01} \tilde{B}_d (P^{11})'] G^- \begin{bmatrix} P^{10} \tilde{B}_d (P^{01})' \\ P^{11} \tilde{B}_d (P^{01})' \end{bmatrix}, \tag{6.4.3}$$

where $\tilde{B}_d$ is as in Lemma 6.4.1 and G^- is a generalized inverse of a matrix G, with G given by

$$G = \begin{bmatrix} P^{10} \tilde{B}_d (P^{10})' & P^{10} \tilde{B}_d (P^{11})' \\ P^{11} \tilde{B}_d (P^{10})' & P^{11} \tilde{B}_d (P^{11})' \end{bmatrix}.$$

Using the matrix A_d as above, Bose and Dey (2003) obtained the following result.

Theorem 6.4.1. *Under the non-additive model (6.4.1) with random subject effects, a design $d^* \in \Omega_{t,n,p}$, which is strongly balanced, uniform on the periods and uniform on the subjects in the first $(p-1)$ periods is universally optimal for the estimation of carryover effects over $\Omega_{t,n,p}$.*

Proof. For any design $d \in \Omega_{t,n,p}$, by (6.4.3),

$$A_d \leq P^{01}\tilde{B}_d(P^{01})',$$

in the Loewner sense. Also, by Lemma 6.4.1, $\tilde{B}_d \leq \sigma^{-2}\tilde{V}_d$. These together imply that

$$A_d \leq \sigma^{-2}P^{01}\tilde{V}_d(P^{01})'. \tag{6.4.4}$$

But from (6.4.2) and the expression for $\tilde{V}_d$ as shown in (6.2.7),

$$\text{tr}\left(P^{01}\tilde{V}_d(P^{01})'\right) = \frac{n}{t^2}(p-1)(t-1),$$

for every $d \in \Omega_{t,n,p}$. Hence by (6.4.4),

$$\text{tr}(A_d) \leq \frac{n}{\sigma^2 t^2}(p-1)(t-1), \tag{6.4.5}$$

for every $d \in \Omega_{t,n,p}$.

On the other hand, using the properties of d^*, one can check that

$$P^{01}\tilde{V}_{d^*}(P^{01})' = \frac{n}{t^2}(p-1)I_{t-1},$$

$$P^{01}\tilde{M}_{d^*} = \mathbf{0},\;\; P^{01}\tilde{N}_{d^*}H_n\tilde{N}'_{d^*} = \mathbf{0},\;\; P^{01}\tilde{B}_{d^*}(P^{10})' = \mathbf{0},\;\; P^{01}\tilde{B}_{d^*}(P^{11})' = \mathbf{0}.$$

Hence by (6.4.3) and Lemma 6.4.1,

$$A_{d^*} = \frac{n}{\sigma^2 t^2}(p-1)I_{t-1}. \tag{6.4.6}$$

By (6.4.6), A_{d^*} is completely symmetric and its trace equals the upper bound in (6.4.5). Hence the result follows on applying Theorem 1.4.1. □

Theorem 6.4.1 shows that Theorem 6.3.4 remains robust if the subject effects are random, which implies that the result of Theorem 2.4.2 for carryover effects remains robust under a non-additive model with random subject effects.

6.5 Optimality in the Presence of Higher Order Carryover Effects and Interactions

Several authors, e.g., John and Quenouille (1977) and Collombier and Merchermek (1993), considered models for the analysis of a crossover experiment in which carryover effects persisting beyond one successive period were incorporated. Bose and Mukherjee (2000) studied the optimality of crossover designs under such models.

Following the terminology of John and Quenouille (1977, p. 198), the carryover effect occurring at the rth period following the period of application is called the rth order carryover effect, $r = 1, 2, \ldots$. The model that we consider in this section incorporates into the basic model (1.2.1) all carryover effects up to the kth order and interactions of all orders among treatments applied in successive periods, up to a maximum of k consecutive periods. This model can be specified as

$$\begin{aligned} Y_{ij} &= \mu + \alpha_i + \beta_j + \psi_{d(i,j),d(i-1,j),\ldots,d(i-k,j)} + \epsilon_{ij}, \\ &\qquad\qquad k+1 \le i \le p, \quad 1 \le j \le n, \\ &= \mu + \alpha_i + \beta_j + \psi_{d(i,j),d(i-1,j),\ldots,d(1,j)} + \epsilon_{ij}, \\ &\qquad\qquad 1 \le i \le k, \quad 1 \le j \le n, \end{aligned} \tag{6.5.1}$$

where, for instance, $\psi_{h_1,h_2,\ldots,h_{k+1}}$ is the effect produced when treatment h_1 is applied in the current period, h_2 in the immediately preceding period, $\ldots, h_{k+1}$ in the kth preceding period, $1 \le h_1, h_2, \ldots, h_{k+1} \le t$; all other terms are as in model (1.2.1). Thus $\psi_{d(i,j),d(i-1,j),\ldots,d(i-s,j)}$ stands for the sum of the direct effect of $d(i,j)$, the first order carryover effect of $d(i-1,j), \ldots,$ the sth order carryover effect of $d(i-s,j)$ and all interaction effects between these $s+1$ treatments.

Model (6.5.1) is a generalization of model (6.2.1). Here, k is any positive integer. An experimenter may choose an appropriate value of k, say 1, 2 or 3 etc., depending on her/his belief as to the order of carryovers which might affect the response at a given period. For $k = 1$, we only allow for the first order residual effect. Similar models for $k = 3$ were used by John and Quenouille (1977) for the analysis of data from a crossover experiment.

A correspondence between a t treatment crossover experiment and a t^2 factorial was used in Section 6.2. A similar approach is used here for studying the optimality of crossover designs under model (6.5.1). We consider a t^{k+1} factorial experiment with treatment combinations represented as $(h_1, h_2, \ldots, h_{k+1})$ where $1 \le h_1, h_2, \ldots, h_{k+1} \le t$, such that h_1 represents the treatment producing the direct effect and h_i, $2 \le i \le k+1$, represents the treatment producing the $(i-1)$th order carryover effect on a subject. Then the properties of a design in $\Omega_{t,n,p}$ can be studied under the model (6.5.1) by looking upon this design as a symmetric t^{k+1} factorial experiment where the direct and the $(i-1)$th order carryover effects $(2 \leq i \leq k+1)$ may be interpreted as the main effects of factors F_1 and F_i, $2 \leq i \leq k+1$, respectively, and the interactions between the treatments in the $k+1$ successive periods are given by the corresponding factorial interaction effects.

Analogous to (6.2.2), writing the $t^{k+1} \times 1$ vector $\boldsymbol{\xi}$ as

$$\boldsymbol{\xi} = (\xi_{11\ldots1}, \xi_{11\ldots12}, \ldots, \xi_{11\ldots1t}, \ldots, \xi_{t1\ldots1}, \ldots, \xi_{tt\ldots t})', \tag{6.5.2}$$

we may rewrite (6.5.1) in the form

$$Y_{ij} = \mu + \alpha_i + \beta_j + \boldsymbol{\lambda}'_{ij}\boldsymbol{\xi} + \epsilon_{ij}, \ \ 1 \leq i \leq p, \ 1 \leq j \leq n,$$

where now the $\boldsymbol{\lambda}_{ij}$ take the form

$$\boldsymbol{\lambda}_{ij} = \begin{cases} \boldsymbol{e}_{d(i,j)} \otimes \boldsymbol{e}_{d(i-1,j)} \otimes \cdots \otimes \boldsymbol{e}_{d(i-k,j)}, \\ \qquad\qquad k+1 \leq i \leq p, 1 \leq j \leq n, \\ \boldsymbol{e}_{d(i,j)} \otimes \boldsymbol{e}_{d(i-1,j)} \otimes \cdots \otimes \boldsymbol{e}_{d(1,j)} \otimes (\prod \otimes\, t^{-1}\mathbf{1}_t), \\ \qquad\qquad 1 \leq i \leq k, 1 \leq j \leq n, \end{cases} \tag{6.5.3}$$

and $\prod \otimes$ denotes the Kronecker product of $(k+1-i)$ terms and $\boldsymbol{e}_{d(i,j)}$ is as in (6.2.4). The analysis now proceeds essentially as in Section 6.2. First, one needs to define $\tilde{V}_d$, $\tilde{M}_d$ and $\tilde{N}_d$ via (6.2.7), now using the $\boldsymbol{\lambda}_{ij}$ as shown in (6.5.3). Then the coefficient matrix of the reduced normal equations for estimating $\boldsymbol{\xi}$ is obtained from (6.2.8).

We can now extend the definition of strong balance to this situation as follows.

Definition 6.5.1. Under the model (6.5.1), a design $d \in \Omega_{t,n,p}$ is called strongly balanced of order k if each consecutive subset of i periods in d contains each i-tuple of treatments equally often, $1 \leq i \leq k+1$.

Let $d^* \in \Omega_{t,n,p}$ be a uniform strongly balanced design of order k. Then we have the following result due to Bose and Mukherjee (2000).

Theorem 6.5.1. *Under model (6.5.1), d^* is universally optimal for the estimation of direct effects over $\Omega_{t,n,p}$.* □

The proof of this theorem rests essentially on the fact that under d^*, contrasts belonging to direct effects are estimable orthogonally to those belonging to carryover effects of different orders and also to those belonging to the various interactions.

Remark 6.5.1. The model (6.5.1) and the designs d^* may be useful in situations where the successive periods are close to each other in time and there is not enough time lag between them for the carryover effect to die out just after one period. However, since model (6.5.1) is rather general, the optimal designs under this model are larger in size compared to those under the models incorporating first order carryover effects only. As expected, the number of subjects increases with increase in the value of k so that, the smaller the value of k, the smaller is the optimal design. □

Two examples of the optimal designs of Theorem 6.5.1 are given below from Bose and Mukherjee (2000).

Example 6.5.1. Suppose $k = 2, t = 3$. The design d_1 shown below is universally optimal for direct effects over $\Omega_{3,27,9}$ under a non-additive model with up to second order carryover effects.

$$d_1 = \begin{array}{ccccccccc}
000 & 000 & 000 & 111 & 111 & 111 & 222 & 222 & 222 \\
012 & 012 & 012 & 012 & 012 & 012 & 012 & 012 & 012 \\
012 & 120 & 201 & 012 & 120 & 201 & 012 & 120 & 201 \\
111 & 111 & 111 & 222 & 222 & 222 & 000 & 000 & 000 \\
120 & 120 & 120 & 120 & 120 & 120 & 120 & 120 & 120 \\
120 & 201 & 012 & 120 & 201 & 012 & 120 & 201 & 012 \\
222 & 222 & 222 & 000 & 000 & 000 & 111 & 111 & 111 \\
201 & 201 & 201 & 201 & 201 & 201 & 201 & 201 & 201 \\
201 & 012 & 120 & 201 & 012 & 120 & 201 & 012 & 120
\end{array}.$$

□

Example 6.5.2. Suppose $k = 3, t = 2$. The design d_2 shown below is universally optimal for direct effects over $\Omega_{2,16,8}$.

$$d_2 = \begin{array}{c}
0000000011111111 \\
0101010101010101 \\
0101101001011010 \\
0110100101101001 \\
1111111100000000 \\
1010101010101010 \\
1010010110100101 \\
1001011010010110
\end{array}.$$

□

The models in Sections 6.2, 6.4 and 6.5 relax one or more of the assumptions of model (1.2.1), for example, (6.2.1) relaxes the assumption of additivity of treatment effects in successive periods, (6.5.1) allows higher order carryovers, while the model in Section 6.4 allows random subject effects. Combining all of them, we can as well consider a single model as

$$Y_{ij} = \begin{cases} \mu + \alpha_i + \beta_j + \psi_{d(i,j),d(i-1,j),\ldots,d(i-k,j)} + \epsilon_{ij}, \\ \qquad\qquad k+1 \leq i \leq p, \quad 1 \leq j \leq n, \\ \mu + \alpha_i + \beta_j + \psi_{d(i,j),d(i-1,j),\ldots,d(1,j)} + \epsilon_{ij}, \\ \qquad\qquad 1 \leq i \leq k, \quad 1 \leq j \leq n, \end{cases} \tag{6.5.4}$$

where the only change from (6.5.1) is that in (6.5.4), β_j, the effect of subject j, $1 \leq j \leq n$, is a random variable. We assume, as in Section 6.4, that $\boldsymbol{\beta} = (\beta_1, \ldots, \beta_n)'$ and the error components are independently distributed of each other with zero means and dispersion matrices $\sigma_\beta^2 I_n$ and $\sigma^2 I_{np}$, respectively. The resultant structure is such that observations from the same subject are correlated while those arising from different subjects are uncorrelated. Note that if we now take $k = 1$ and treat the subject effects as fixed, we get back the model (1.2.1).

Using the Kronecker calculus for factorial arrangements as before, Bose and Mukherjee (2003) proved the following result.

Theorem 6.5.2. *Under the model (6.5.4), a uniform design $d^* \in \Omega_{t,n,p}$ which is also strongly balanced of order k, is universally optimal for the estimation of direct effects over $\Omega_{t,n,p}$.* □

For some more results under a mixed effects non-additive model incorporating random subject effects, see Bose and Dey (2006).

Chapter 7

Some Further Developments

7.1 Introduction

In this chapter, we discuss some other developments in the context of optimal crossover designs which were not covered in the earlier chapters. We begin with designs involving just two treatments. Two-treatment crossover designs are popular in the context of clinical trials for comparing a pair of treatments, typically, a new drug with a standard one or a treatment with a placebo. Several authors have obtained results on optimality of designs with two treatments, under models with uncorrelated, and also correlated, error structures. Some of these results which are obtained via the approximate design theory are reviewed in Section 7.2. In Section 7.3 we consider models with correlated error structures and present optimality results in exact design theory under such models for arbitrary t.

Next we consider a problem that often arises in practice when one wants to determine the relative performance of a number of *test treatments* vis-a-vis a standard treatment, called the *control.* For instance, in many pharmaceutical studies the experimenter is interested in comparing some new drugs (test treatments) with an established standard drug, which acts as the control treatment. In the context of incomplete block designs, the problem of finding optimal designs for test-control treatment comparisons has received considerable attention in the literature and for authoritative reviews and additional references, we refer to Hedayat, Jacroux and Majumdar (1988) and Majumdar (1996). For crossover designs, the problem of finding optimal designs for test-control treatment comparisons has been considered by several authors in the recent past and these results are reviewed in Section 7.4.

The last problem we consider in this chapter concerns that of subject dropout during experimentation. In certain applications, sometimes it may not be possible to carry out a planned crossover trial in its entirety. For example, in the context of clinical trials with patients as subjects, it is often seen that certain patients drop out before the experiment is over. In such cases the resulting design after dropouts may turn out to be inefficient or sometimes, even disconnected. It is therefore important to study designs that remain optimal when the trial is truncated prematurely. This problem has received attention in recent years. In Section 7.5, some aspects of this problem are discussed. Finally, in Section 7.6, additional comments regarding some issues related to crossover designs are made.

7.2 Optimal Two-treatment Designs

When there are only two treatments, a universally optimal design for direct effects over $\Omega_{2,n,p}$ is one that minimizes the variance of the best linear unbiased estimator (BLUE) of $\tau_1 - \tau_2$ over all designs in $\Omega_{2,n,p}$, where as before, τ_s is the direct effect of treatment s, $s = 1, 2$. Similarly, when carryover effects are of interest, one would like to minimize the variance of the BLUE of $\rho_1 - \rho_2$ over $\Omega_{2,n,p}$.

The optimality and efficiency of two-treatment crossover designs have been studied by many authors and under various modifications of the traditional model (1.3.3). Kershner and Federer (1981) assumed a model with sequence effects instead of subject effects and gave the variances of unbiased estimators of direct effect contrasts, while Laska and Meisner (1985) derived optimal designs under a model with random subject effects and equicorrelated covariance structure for general p, and also with an autoregressive covariance structure for $p = 2, 3$ and 4. Adopting model (1.3.14) with autoregressive errors, Matthews (1987) gave optimal and efficient designs for $p = 3$ and 4. Results for general p were obtained under the uncorrelated error model (1.3.3) by Matthews (1990), the correlated error model (1.3.14) with autoregressive errors by Kunert (1991), and model (1.3.14) with a general form of V by Kushner (1997a). For some additional results and references on efficient two-treatment crossover designs, we refer to Laska, Meisner and Kushner (1983), Ebbutt (1984), Carriere and Reinsel (1992) and Carriere and Huang (2000). In this section we review some of the results by the above authors.

7.2.1 *Designs under Uncorrelated Errors*

In the context of designs with two treatments, Laska and Meisner (1985) introduced the notion of *dual balanced designs.* These designs allocate the same number of subjects to a treatment sequence $\boldsymbol{s}$ and its dual sequence, the latter being obtained by interchanging treatment labels 1 and 2 in $\boldsymbol{s}$. Clearly, such dual balanced designs are symmetric designs in the sense of Definition 4.2.2, but restricted to $t = 2$. Consequently, their optimality properties follow from the results on optimality of symmetric designs as given in Chapter 4 for arbitrary $t(\geq 2)$ (see Examples 4.6.1 and 4.7.2 of Chapter 4).

For the uncorrelated error model (i.e., model (1.3.3)), Chapter 4 gives a detailed discussion on optimality via the approximate design theory approach for general t. In view of this, we do not review here the results of Matthews (1990) and others which are also proved under this model but specialized to $t = 2$. However, for the sake of completeness, we give some illustrative examples of their optimal designs. It can be easily verified that the designs in these examples satisfy the conditions of Theorem 4.6.2 and/or Theorem 4.7.1, thus establishing their optimality.

Example 7.2.1. Let $p = 3$. Under model (1.3.3) with $n = 2$, the design shown below is universally optimal for both direct and carryover effects over $\Omega_{2,2,3}$.

$$\begin{matrix} 1 & 2 \\ 2 & 1 \\ 2 & 1 \end{matrix}.$$

□

Example 7.2.2. Let $p = 4$. In this case (as also in several other cases, as seen in Chapter 4) a universally optimal design can be obtained for various values of n and more than one non-isomorphic optimal design can be found for the same n. Designs d_1, d_2, d_3 shown below are universally optimal for both direct and carryover effects in $\Omega_{2,n,4}$ with $n = 4, 8$ and 12, respectively. Again, by juxtaposing d_1 twice, we can obtain another universally optimal design in $\Omega_{2,8,4}$ which is different from d_2. Similarly juxtaposing d_1 thrice, we get an optimal design in $\Omega_{2,12,4}$ different from d_3.

$$d_1 \equiv \begin{matrix} 1\,2\,1\,2 \\ 1\,2\,2\,1 \\ 2\,1\,2\,1 \\ 2\,1\,1\,2 \end{matrix}, \quad d_2 \equiv \begin{matrix} 1\,1\,1\,2\,2\,2\,1\,2 \\ 1\,1\,1\,2\,2\,2\,2\,1 \\ 2\,2\,2\,1\,1\,1\,1\,2 \\ 2\,2\,2\,1\,1\,1\,2\,1 \end{matrix},$$

$$d_3 \equiv \begin{matrix} 1\,1\,1\,1\,2\,2\,2\,2\,1\,2\,1\,2 \\ 1\,1\,1\,1\,2\,2\,2\,2\,2\,1\,2\,1 \\ 2\,2\,2\,2\,1\,1\,1\,1\,1\,2\,2\,1 \\ 2\,2\,2\,2\,1\,1\,1\,1\,2\,1\,1\,2 \end{matrix}.$$

□

For the correlated error model (i.e., model (1.3.14)), however, optimality results for general t were not discussed in detail in Chapter 4. In the next subsection, we review some of these results for $t = 2$.

7.2.2 *Designs under Correlated Errors*

We consider model (1.3.14) where V has a general form and p is arbitrary. Under this model, we first follow the approach of Chapter 4 adapted to the case of $t = 2$. Following Kushner (1997a), we use the approximate design theory to identify optimal designs in a general class. We use the notation from Chapter 4 but the algebra is simpler here as there are only two treatments. Then we specialize to the case when V is autoregressive and illustrate that this method can easily be applied to obtain optimal designs for general p, including the ones given in Matthews (1987) for $p = 3$ and 4. Similarly, this can lead to an extension of some of the results of Laska and Meisner (1985) to general p. Subsequently, following Kunert (1991) we make some more comments.

As in Chapter 4, here we start with the class of all 2^p sequences $\mathcal{S}$, where each sequence is a $p \times 1$ vector, with elements being either 1 or 2. From Section 4.2, recall that $T_{\boldsymbol{s}}$ and $F_{\boldsymbol{s}}$ are the $p \times t$ incidence matrices for the period versus direct effects and the period versus carryover effects, respectively, for each $\boldsymbol{s} \in \mathcal{S}$. These matrices for $t = 2$ may be written in partitioned form as

$$T_{\boldsymbol{s}} = (T_{\boldsymbol{s},1} \ \ T_{\boldsymbol{s},2}) \quad \text{and} \quad F_{\boldsymbol{s}} = (F_{\boldsymbol{s},1} \ \ F_{\boldsymbol{s},2}),$$

where $T_{\boldsymbol{s},m}$ and $F_{\boldsymbol{s},m}$ are the columns corresponding to treatment m in $T_{\boldsymbol{s}}$ and $F_{\boldsymbol{s}}$, respectively, $m = 1, 2$. Then the quadratic in (4.3.1) may be considerably simplified as shown in the following lemma. For easy reference, we recall from (1.3.17) that

$$V^* = V^{-1} - (V^{-1} J_p V^{-1})/(\mathbf{1}_p' V^{-1} \mathbf{1}_p). \tag{7.2.1}$$

Lemma 7.2.1. *With $t = 2$ and general V, the quadratic (4.3.1) is given by*

$$q_{\boldsymbol{s}}(u) = q_{11}^{\boldsymbol{s}} + 2q_{12}^{\boldsymbol{s}}u + q_{22}^{\boldsymbol{s}}u^2, \ -\infty < u < \infty, \tag{7.2.2}$$

where

$$q_{11}^{\boldsymbol{s}} = \tfrac{1}{2}(2T_{\boldsymbol{s},1} - \mathbf{1}_p)'V^*(2T_{\boldsymbol{s},1} - \mathbf{1}_p)$$

$$q_{12}^{\boldsymbol{s}} = \tfrac{1}{2}(2T_{\boldsymbol{s},1} - \mathbf{1}_p)'V^*(2F_{\boldsymbol{s},1} - \bar{\mathbf{1}}_p)$$

$$q_{22}^{\boldsymbol{s}} = \tfrac{1}{2}(2F_{\boldsymbol{s},1} - \bar{\mathbf{1}}_p)'V^*(2F_{\boldsymbol{s},1} - \bar{\mathbf{1}}_p),$$

with $\bar{\mathbf{1}}_p$ denoting a $p \times 1$ vector with its first element equal to zero and all other elements unity. Furthermore,

$$q'_{\boldsymbol{s}}(u) = (2T_{\boldsymbol{s},1} - \mathbf{1}_p)'V^*(2F_{\boldsymbol{s},1} - \bar{\mathbf{1}}_p) + u(2F_{\boldsymbol{s},1} - \bar{\mathbf{1}}_p)'V^*(2F_{\boldsymbol{s},1} - \bar{\mathbf{1}}_p). \tag{7.2.3}$$

Proof. From (4.3.1)

$$q_{11}^{\boldsymbol{s}} = \operatorname{tr}\left[\{(T_{\boldsymbol{s},1} \ \ T_{\boldsymbol{s},2}) - \tfrac{1}{2}(\mathbf{1}_p \ \ \mathbf{1}_p)\}'V^*\{(T_{\boldsymbol{s},1} \ \ T_{\boldsymbol{s},2}) - \tfrac{1}{2}(\mathbf{1}_p \ \ \mathbf{1}_p)\}\right]$$

$$= (1/4) \sum_{m=1}^{2} (2T_{\boldsymbol{s},m} - \mathbf{1}_p)'V^*(2T_{\boldsymbol{s},m} - \mathbf{1}_p)$$

$$= (1/2)\,(2T_{\boldsymbol{s},1} - \mathbf{1}_p)'V^*(2T_{\boldsymbol{s},1} - \mathbf{1}_p)\,,$$

on observing that for any sequence $\boldsymbol{s}$, $T_{\boldsymbol{s},2} = \mathbf{1}_p - T_{\boldsymbol{s},1}$. Similarly, and using the fact that $F_{\boldsymbol{s},2} = \bar{\mathbf{1}}_p - F_{\boldsymbol{s},1}$, the expressions for the other coefficients may be derived. □

It is clear that the quadratic in (7.2.2) remains the same for a sequence $\boldsymbol{s}$ and also its dual (or equivalent) sequence which together constitute the symmetry block (or equivalence class) $< \boldsymbol{s} >$. So, in the context of optimality, it is enough to consider only the distinct symmetry blocks $< \boldsymbol{s} >$ in $\mathcal{S}$. Indeed, as in Chapter 4, we may restrict to dual balanced (i.e., symmetric) designs while searching for optimal designs. Now, using the simplified form (7.2.2) of $q_{\boldsymbol{s}}(u)$, one can employ Lemma 4.4.2 to find b, a and $\bar{\mathcal{S}}$ as introduced in (4.3.6). With these b, a and $\bar{\mathcal{S}}$, Kushner (1997a) proved the following results for general V and $t = 2$. Recall that similar results for $V = I_p$ and general t were obtained in Theorems 4.5.1 and 4.6.1.

Theorem 7.2.1. *For $t = 2$, the information matrix for direct effects of a universally optimal design is given by*

$$C_d = nbH_2 = \frac{nb}{2}\begin{pmatrix} 1 & -1 \\ -1 & 1 \end{pmatrix},$$

where b is as obtained from (4.3.6) with $q_{\boldsymbol{s}}(u)$ as in (7.2.2). □

Theorem 7.2.2. *For $t = 2$, a dual balanced design is universally optimal for direct effects if and only if its support is in $\bar{S}$ and the condition*

$$\sum_{\boldsymbol{s}\in\bar{S}} P_{\boldsymbol{s}} q'_{\boldsymbol{s}}(a) = 0$$

holds, where $P_{\boldsymbol{s}}$ is the weight of the symmetry block $< \boldsymbol{s} >$, a, $\bar{S}$ are as in (4.3.6) with $q'_{\boldsymbol{s}}(a)$ as in (7.2.3). □

In order to apply the above theorems we need to know a, b and $\bar{S}$ and these may be determined using Lemma 4.4.2 and (7.2.2) provided V^* is known. Also, the efficiency of a dual balanced design $d \leftrightarrow P^*$ may be computed, essentially along the line of Section 4.8 in Chapter 4. For example, the efficiency (Eff_d) for direct effects is given by

$$\text{Eff}_d = \frac{q_{11}(P^*) - q_{12}^2(P^*)/q_{22}(P^*)}{b}.$$

Using the above technique under an autoregressive error structure, optimal designs obtained by Matthews (1987) for $p = 3$ and 4 can be seen to be optimal. We consider results under such an error structure in the next subsection.

7.2.3 *Designs under Autoregressive Errors*

While studying designs under a correlated error model, it is often assumed that the errors for each subject follow a stationary first order autoregressive process. In the context of crossover designs this assumption is quite plausible since if measurements occur at equidistant periods in time then under this assumption, the correlation between errors on the same subject decreases exponentially with time. Under such an error structure, crossover designs have been studied by several authors, including Azzalini and Giovagnoli (1987), Berenblut and Webb (1974), Bora (1984, 1985), Kunert (1985, 1991), Laska and Meisner (1985), Matthews (1987), Gill and Shukla (1987), Carriere and Reinsel (1992) and Carriere and Huang (2000). The problem of finding optimal designs under such a model with $t = 2$ has been considered by Laska and Meisner (1985), Matthews (1987), Kunert (1991) and Kushner (1997a). In this subsection, we review some of these results.

We now assume model (1.3.14) with the error dispersion matrix

$$\sigma^2\Sigma = \sigma^2 I_n \otimes V,$$

where $V = (v_{ii'})$, with

$$v_{ii'} = \frac{\rho^{|i-i'|}}{1-\rho^2}, \; 1 \le i, i' \le p, \tag{7.2.4}$$

and ρ is a constant such that $-1 < \rho < 1$. The value of the correlation coefficient ρ will often not be known in practice though as Kunert (1985) argues, it may be possible to get some information about the value of ρ from similar experiments already analyzed. In the following discussion while searching for optimal designs we assume that ρ is known. Incidentally, while the form (7.2.4) has been commonly used in the literature in the context of autoregressive errors, some authors have also used a variant thereof with the divisor $(1-\rho^2)$ omitted.

An advantage of the autoregressive error structure is that under such a V, the matrix V^{-1} has the following simple form

$$V^{-1} = \begin{pmatrix} 1 & -\rho & 0 & \cdots & 0 & 0 & 0 \\ -\rho & 1+\rho^2 & -\rho & \cdots & 0 & 0 & 0 \\ 0 & -\rho & 1+\rho^2 & \cdots & 0 & 0 & 0 \\ \vdots & \vdots & \vdots & \vdots & \vdots & \vdots & \vdots \\ 0 & 0 & 0 & \cdots & 1+\rho^2 & -\rho & 0 \\ 0 & 0 & 0 & \cdots & -\rho & 1+\rho^2 & -\rho \\ 0 & 0 & 0 & \cdots & 0 & -\rho & 1 \end{pmatrix} \tag{7.2.5}$$

and thus V^* can be computed easily from (7.2.1). Theorem 7.2.2 can then be applied to get optimal designs. The following example illustrates this.

Example 7.2.3. Let $p = 3$. Then from (7.2.4) and (7.2.5)

$$V = \frac{1}{1-\rho^2}\begin{pmatrix} 1 & \rho & \rho^2 \\ \rho & 1 & \rho \\ \rho^2 & \rho & 1 \end{pmatrix} \text{ and } V^{-1} = \begin{pmatrix} 1 & -\rho & 0 \\ -\rho & 1+\rho^2 & -\rho \\ 0 & -\rho & 1 \end{pmatrix}.$$

Hence by (7.2.1), on simplification,

$$V^* = \frac{1}{3-\rho}\begin{pmatrix} 2 & -1-\rho & -1+\rho \\ -1-\rho & 2+2\rho & -1-\rho \\ -1+\rho & -1-\rho & 2 \end{pmatrix}.$$

Here

$$\mathcal{S} = <(111)'> \cup <(112)'> \cup <(121)'> \cup <(122)'>,$$

where as before, $< \boldsymbol{s} >$ denotes the symmetry block consisting of the sequence $\boldsymbol{s}$ and its dual (or equivalent sequence). For $\boldsymbol{s} = (111)'$, $T_{\boldsymbol{s},1} =$

$(111)'$ and so, from Lemma 7.2.1,

$$q_{11}^{(111)'} = \frac{1}{2}(1\ 1\ 1)V^* \begin{pmatrix} 1 \\ 1 \\ 1 \end{pmatrix} = 0.$$

Similarly, using $T_{(112)',1} = (110)'$, $T_{(121)',1} = (101)'$, $T_{(122)',1} = (100)'$, the four $q_{11}^{\boldsymbol{s}}$ values are

$$q_{11}^{(111)'} = 0, \quad q_{11}^{(112)'} = \frac{4}{3-\rho}, \quad q_{11}^{(121)'} = \frac{4(1+\rho)}{3-\rho}, \quad q_{11}^{(122)'} = \frac{4}{3-\rho}. \quad (7.2.6)$$

Suppose $\rho < 0$. We take $A = 0$ in Lemma 4.4.2 and use (7.2.6) to get

$$B = \max_{\boldsymbol{s}}\{q_{\boldsymbol{s}}(0)\} = \max_{\boldsymbol{s}}\{q_{11}^{\boldsymbol{s}}\} = \frac{4}{3-\rho},$$

and

$$S_A = \{< (112)' >, \ < (122)' >\}.$$

Moreover, S_A^+ and S_A^- in Lemma 4.4.2 are nonempty here because, from (7.2.3), on simplification we get

$$q'_{(112)'}(0) = \frac{2(\rho-1)}{3-\rho} < 0, \quad q'_{(122)'}(0) = \frac{-4\rho}{3-\rho} > 0.$$

Hence Lemma 4.4.2 yields

$$a = 0, \quad b = \frac{4}{3-\rho}, \qquad \bar{S} = \{< (112)' >, \ < (122)' >\}.$$

Using Theorem 7.2.2, we can now construct a universally optimal design with support on only the two symmetry blocks in $\bar{S}$ such that their weights satisfy

$$2(\rho-1)P_{(112)'} - 4\rho P_{(122)'} = 0, \quad P_{(112)'} + P_{(122)'} = 1.$$

This leads to

$$P_{(112)'} = \frac{2\rho}{3\rho-1}, \quad P_{(122)'} = \frac{\rho-1}{3\rho-1}.$$

So, a universally optimal design allocates each of the sequences $(112)'$ and $(221)'$ to $\frac{\rho}{3\rho-1}$ subjects and each of the sequences $(122)'$ and $(211)'$ to $\frac{\rho-1}{2(3\rho-1)}$ subjects. □

The above optimal design was also obtained by Matthews (1987) by direct methods involving lengthy algebra. Some other optimal designs obtained by him for $p = 3$ and $p = 4$ can also be obtained via the above technique. A similar approach works for constructing optimal designs for carryover effects too. However, when ρ is not known it may be meaningful to look for designs which are efficient over a large range of ρ.

Matthews (1987) provided tables of efficiencies of several simpler dual balanced designs with $p = 3$ and 4 for nine equidistant values of ρ in the range $-0.8 \leq \rho \leq 0.8$. These efficiencies are given for both direct and carryover effects and for each ρ, the efficiencies are computed as the ratio of the variance of the contrast estimator as obtained under the optimum design to that obtained under the given design. For several designs, these efficiencies are quite high. Besides showing how efficient a design is for a particular value of ρ, these tables help one in choosing a simple design which remains efficient over a possible range of values of ρ. The latter point is of practical importance since ρ is typically unknown. From these tables, we give some examples of designs and their efficiencies below. These simple and efficient designs could be useful in practice.

Example 7.2.4. For $p = 3$ and $p = 4$, consider the designs

$$d_1 = \begin{matrix} 1 & 2 & 1 & 2 \\ 2 & 1 & 1 & 2 \\ 2 & 1 & 2 & 1 \end{matrix}, \quad d_2 = \begin{matrix} 1 & 2 & 1 & 2 \\ 2 & 1 & 2 & 1 \\ 2 & 1 & 1 & 2 \end{matrix}, \quad d_3 = \begin{matrix} 1 & 2 \\ 2 & 1 \\ 2 & 1 \end{matrix},$$

$$d_4 = \begin{matrix} 1 & 2 & 1 & 2 \\ 1 & 2 & 2 & 1 \\ 2 & 1 & 2 & 1 \\ 2 & 1 & 1 & 2 \end{matrix}, \quad d_5 = \begin{matrix} 1 & 2 \\ 1 & 2 \\ 2 & 1 \\ 2 & 1 \end{matrix}, \quad d_6 = \begin{matrix} 1 & 2 \\ 2 & 1 \\ 2 & 1 \\ 1 & 2 \end{matrix}.$$

For $p = 3$ and for direct effects, d_1 has an efficiency of at least 90% for ρ in the range -0.8 to 0.8 and its efficiency becomes less than 97% only when $\rho > 0.6$. For carryover effects, d_2 remains efficient, except when ρ is close to 1. The design d_3 is efficient for both direct and carryover effects and these efficiencies fall below 80% only when $\rho < -0.6$. Note that d_3 is the extra period strongly balanced design which is optimal when $\rho = 0$ (see Theorem 2.4.2).

For $p = 4$, d_4 seems to be the best for both direct and carryover effects for all ρ. This is the strongly balanced uniform design which is optimal when $\rho = 0$ (see Theorem 2.4.1). For direct effects, d_5 and d_6 are highly efficient for $\rho < 0$ and $\rho > 0$, respectively. For carryover effects, these roles of d_5 and d_6 are reversed. □

Recall that for a given d, the information matrix for direct effects C_d is as in (1.3.18), now with V as given by (7.2.4). Kunert (1991) considered the information matrix, given by $\tilde{C}_d$, say, under a simpler model with no subject and period effects. Clearly, as in (1.4.6),

$$C_d \leq \tilde{C}_d.$$

Though $\tilde{C}_d$ involves a g-inverse which is difficult to obtain in general, for the case $t = 2$, Kunert (1991) showed that $\tilde{C}_d$ has a particularly simple form. Moreover, for a certain type of designs with $t = 2$, the difference between C_d and $\tilde{C}_d$ becomes negligible for large p. In view of this, Kunert (1991) suggested the maximization of $\mathrm{tr}(\tilde{C}_d)$ instead of $\mathrm{tr}(C_d)$ and called a design which achieves this maximization as "efficient" for direct effects. He further noted that these efficient designs are optimal under a model with no subject effect. Such a model was considered by Laska and Meisner (1985) and some of their optimal designs are indeed these efficient designs. Kunert (1991) identified these efficient designs and gave a method for constructing them for any given p and ρ. We summarize below the key steps of the construction as proposed by him.

Case 1. p even. Define the following sequences:

$$\begin{aligned} \boldsymbol{s}_1 &= (112211\ldots22)' \; \boldsymbol{s}_2 = (12211\ldots221)' \\ \boldsymbol{s}_3 &= (121212\ldots12)', \; \boldsymbol{s}_4 = (11\ldots122\ldots2)', \end{aligned}$$

where $\boldsymbol{s}_4$ contains the treatment 1, $p/2$ times. Treatments 1 and 2 appear equally often in $\boldsymbol{s}_3$ and $\boldsymbol{s}_4$ and, also in $\boldsymbol{s}_1$ and $\boldsymbol{s}_2$ if $p = 0 \bmod 4$. If $p = 2 \bmod 4$, then $\boldsymbol{s}_1$ ends with 11 and $\boldsymbol{s}_2$ with 112. Writing $\psi = |\rho|$, let

$$p_1^* = \frac{1}{2}n(p-2)(\psi + \psi^3),$$

$$p_2^* = \frac{1}{2}n\frac{\{(p-2)(1+\psi^2) + 1 - 2\psi\}(1+\psi)^2 + 2\psi^2(1+\psi)}{2\psi\{(1+\psi)^4 - 4\psi^2\}^{\frac{1}{2}}}$$

$$-\frac{1}{2}n\frac{(p-2)(1+\psi^2) + 1 - 2\psi}{2\psi},$$

and

$$p^* = \begin{cases} p_1^*, & \text{if } p_1^* < n/2, \\ n/2, & \text{if } p_1^* \geq n/2, \quad p_2^* \leq n/2, \\ p_2^*, & \text{otherwise.} \end{cases}$$

To obtain an efficient design we need to follow the two steps given below.

Step 1: When $\rho \geq 0$,

(a) if $p^* \leq n/2$, assign $\boldsymbol{s}_1$ to $n/2 - p^*$ subjects and $\boldsymbol{s}_2$ to $n/2 + p^*$ subjects

(b) if $p^* > n/2$, assign $\boldsymbol{s}_2$ to $n - (2p^* - n)/(p-2)$ subjects and $\boldsymbol{s}_3$ to $(2p^* - n)/(p-2)$ subjects,

and when $\rho < 0$,

(a) if $p^* \leq n/2$, assign $\boldsymbol{s}_1$ to $n/2 + p^*$ subjects and $\boldsymbol{s}_2$ to $n/2 - p^*$ subjects

(b) if $p^* > n/2$, assign $\boldsymbol{s}_1$ to $n - (2p^* - n)/(p-4)$ subjects and $\boldsymbol{s}_4$ to $(2p^* - n)/(p-4)$ subjects.

Step 2: After allocation as in Step 1, replace each sequence by its dual sequence in half of the subjects to which it is assigned.

Case 2. p odd. We indicate the construction for $p = 1 \bmod 4$ (see Kunert (1991) for the case $p = 3 \bmod 4$). Define the following sequences:

$$\begin{aligned} \boldsymbol{s}_5 &= (1122\ldots11221)', & \boldsymbol{s}_6 &= (121122\ldots112)', \\ \boldsymbol{s}_7 &= (1212\ldots12121)', & \boldsymbol{s}_8 &= (12211\ldots2211)' \\ \boldsymbol{s}_9 &= (11122\ldots1122)' & \boldsymbol{s}_{10} &= (11\ldots122\ldots2)', \end{aligned}$$

where treatment 1 occurs $(p-1)/2$ times in $\boldsymbol{s}_{10}$. With p_2^* and ψ as in Case 1, let

$$p_3^* = \frac{1}{2}n\frac{\{(p-2)(1+\psi^2)+1-2\psi\}(1+\psi+\psi^2)-2\psi^2(1-\psi)}{2\psi\{(1+\psi+\psi^2)^2-4\psi^2\}^{\frac{1}{2}}}$$
$$-\frac{1}{2}n\frac{(p-2)(1+\psi^2)+1-2\psi}{2\psi},$$

$$p^* = \begin{cases} p_3^*, \text{ if } p_3^* < n, \\ n, \text{ if } p_3^* \geq n, \ p_2^* \leq n, \\ p_2^*, \text{ otherwise.} \end{cases}$$

As in Case 1, we follow the two steps given below.

Step 1: When $\rho \geq 0$,

(a) if $p^* \leq n$, assign $\boldsymbol{s}_5$ to $n - p^*$ subjects and $\boldsymbol{s}_6$ to p^* subjects

(b) if $p^* > n$, assign $\boldsymbol{s}_6$ to $n - 2(p^* - n)/(p-3)$ subjects and $\boldsymbol{s}_7$ to $2(p^* - n)/(p-3)$ subjects,

and when $\rho < 0$,

(a) if $p^* \leq n$, assign $\boldsymbol{s}_8$ to $n - p^*$ subjects and $\boldsymbol{s}_9$ to p^* subjects

(b) if $p^* > n$, assign s_9 to $n - 2(p^* - n)/(p - 5)$ subjects and s_{10} to $2(p^* - n)/(p - 5)$ subjects.

Step 2: After allocation as in Step 1, replace each sequence by its dual sequence in half of the subjects to which it is assigned.

The designs obtained after Step 2 above in both Cases 1 and 2 are efficient in the sense of Kunert (1991).

Example 7.2.5. Let $p = 8$. After Step 1 as given above, for $\rho \geq 0$, the design d consists of the sequences

11221122	assigned to $n/2 - p^*$ subjects and	
12211221	assigned to $n/2 + p^*$ subjects,	if $p^* \leq n/2$,
12211221	assigned to $7n/6 - p^*/3$ subjects and	
12121212	assigned to $p^*/3 - n/6$ subjects,	if $p^* > n/2$.

On the other hand, for $\rho < 0$, after Step 1, d consists of

11221122	assigned to $n/2 + p^*$ subjects and	
12211221	assigned to $n/2 - p^*$ subjects,	if $p^* \leq n/2$,
11221122	assigned to $5n/4 - p^*/2$ subjects and	
11112222	assigned to $p^*/2 - n/4$ subjects,	if $p^* > n/2$.

Then from the above d, after implementing Step 2, the efficient design may be obtained. Kunert (1991) remarked that these designs have an efficiency greater than 0.995 for all ψ.

Thus these efficient designs essentially consist of two pairs of dual sequences, each of which is allocated to a specified number of subjects. The allocation numbers obtained as above may not be integers and Kunert (1991) recommended that in practice, the nearest integers be used. He also tabulated the efficiencies of these designs for $p = 3, 4, 5$ and $\psi = 0.1, \ldots, 0.9$, and observed that these efficiencies are impressive over the entire range of ψ, specially for $p > 3$.

7.3 Optimal Designs under Correlated Errors for an Arbitrary Number of Treatments

The approximate design theory approach for obtaining optimal designs under model (1.3.14) for arbitrary t and p was discussed in Chapter 4. Alternatively, the exact design theory can also be used to obtain optimal designs in this context and several authors have obtained results in this area. We review a selection of these results in this section and give examples of designs recommended by various authors.

Kunert (1985) considered a version of model (1.3.14) where there are no carryover effects and in which the errors follow a first order autoregressive process with V as in (7.2.4) and ρ known. He studied designs for the case $p = t$ and $n = \mu_1 t$ for some integer $\mu_1 \geq 1$. After defining a function (ρ^*) of t, he showed that a certain type of Williams design is universally optimal for direct effects over the general class $\Omega_{t,n,t}$ whenever $\rho \geq \rho^*$. Interestingly, ρ^* is always negative and equals -1 for $t = 3$. Hence the above mentioned designs are universally optimal whenever $\rho > 0$ and also universally optimal when $t = 3$ for every ρ. Moreover, if we restrict to the class of designs in $\Omega_{t,n,t}$ which are uniform on the subjects, then, for every ρ, these designs are universally optimal over this class.

Gill (1992) considered model (1.3.14) when the errors corresponding to the observations on the same subject follow a first order autoregressive process and obtained conditions under which a design which is uniform on periods and binary over subjects is variance-balanced. He gave two examples of such variance-balanced designs, both of which are orthogonal arrays of type I and strength two (see Definition 3.1.1). A variance-balanced crossover design allows the estimation of all elementary contrasts of direct effects with the same variance, and all elementary contrasts of carryover effects are also estimable with a constant variance. These designs are attractive to experimenters as their analysis is simple. However, an optimal design need not be always variance-balanced; see e.g., Donev and Jones (1995), who gave an example of an A-optimal design that is *not* variance-balanced.

Martin and Eccleston (1998) showed that variance balanced designs can be obtained from orthogonal arrays of type I and made an interesting observation regarding the designs such obtained. This is stated as Lemma 7.3.1 below.

Lemma 7.3.1. *Let $d^* \in \Omega_{t,n,p}$ be a design obtained from an orthogonal array of type I, $OA_I(n, p', t, 2)$, $p' \leq p$, with rows of the array representing the periods (with some rows repeated if $p' < p$), and the columns representing the subjects of d^*. Then under the model (1.3.14), for this design d^*, the matrices C_{duv}, $u, v = 1, 2$, as in (1.3.16) are completely symmetric and of the form $C_{d^*uv} = a_{uv} H_t$, $u, v = 1, 2$, where the a_{uv} depend on which rows of the array are repeated and where they are placed. Here, as usual, $H_t = I_t - J_t/t$.* □

Clearly, with d^* as in Lemma 7.3.1, the information matrices C_{d^*} and $\bar{C}_{d^*}$ are completely symmetric. Hence d^* is variance balanced for both direct and carryover effects. Note that d^* can be binary over subjects (i.e., each subject receives a treatment at most once and $p \leq t$) or non-binary. Martin and Eccleston (1998) gave several examples of these designs which are in a sense optimal for direct and/or carryover effects under specific correlated error structures. One such example is presented below.

Example 7.3.1. Let $t = 3, p = 4$ and $n = 6$. The type I orthogonal array $OA_I(6, 3, 3, 2)$ shown in Example 3.1.1, with any one of its rows repeated once, can be used as a crossover design in $\Omega_{3,6,4}$. There are six possible such designs, of which three are shown below:

$$d_1 = \begin{matrix} 1\,1\,2\,2\,3\,3 \\ 2\,3\,1\,3\,1\,2 \\ 3\,2\,3\,1\,2\,1 \\ 3\,2\,3\,1\,2\,1 \end{matrix}, \quad d_2 = \begin{matrix} 1\,1\,2\,2\,3\,3 \\ 2\,3\,1\,3\,1\,2 \\ 3\,2\,3\,1\,2\,1 \\ 1\,1\,2\,2\,3\,3 \end{matrix}, \quad d_3 = \begin{matrix} 1\,1\,2\,2\,3\,3 \\ 2\,3\,1\,3\,1\,2 \\ 2\,3\,1\,3\,1\,2 \\ 3\,2\,3\,1\,2\,1 \end{matrix}.$$

Design d_1 is an extra-period strongly balanced design which has strong optimality properties under model (1.3.3) (see Theorem 2.4.2). Under model (1.3.14) with $v_{ii'} = ((\rho^{|i-i'|}))$, $1 \leq i, i' \leq p$, Martin and Eccleston (1998) showed that d_1 is optimal for direct effects when $\rho \leq 0.13$, d_2 is optimal when $\rho > 0.64$ while d_3 is optimal when $0.13 < \rho < 0.64$. For carryover effects, d_1 is optimal for all $\rho > 0$. Similar ranges may be obtained for other designs. □

The universal optimality of the OA_I designs under model (1.3.14) with a general structure of V was subsequently proved by Kunert and Martin (2000b) with reference to the class of designs which are binary over subjects. Their result is given below. As in Section 3.4, we write $\mathcal{B}_{t,n,p}$ to denote the class of designs in $\Omega_{t,n,p}$ which are binary over subjects.

Theorem 7.3.1. *For $t \geq p > 2$, let $d^* \in \Omega_{t,n,p}$ be a crossover design given by an orthogonal array of type I, $OA_I(n, p, t, 2)$, with rows of the array representing the periods and the columns representing the subjects of d^*. Then, under the model (1.3.14) where V is any known positive definite matrix, d^* is universally optimal for the estimation of direct effects over $\mathcal{B}_{t,n,p}$.* □

We do not present a detailed proof of this theorem but only give a flavor of their arguments. For this, we write as usual, C_d for the information matrix for direct effects under the model (1.3.14), and denote by $\tilde{C}_d$ the corresponding information matrix under a simpler version of (1.3.14) that excludes period effects. Following (1.4.6), then

$$C_d \leq \tilde{C}_d, \tag{7.3.1}$$

and hence

$$\text{tr}(C_d) \leq \text{tr}(\tilde{C}_d). \tag{7.3.2}$$

Now, with d^* as stated in Theorem 7.3.1, observe that (i) $C_{d^*} = \tilde{C}_{d^*}$, since d^* is uniform over periods, and (ii) by Lemma 7.3.1, C_{d^*} is completely symmetric. However, in general, it is difficult to find an upper bound for $\text{tr}(C_d)$ from (7.3.2) because maximization of $\text{tr}(\tilde{C}_d)$ over all $d \in \Omega_{t,n,p}$ can be challenging. Kunert and Martin (2000b) overcame this difficulty by restricting to the class of binary designs in $\Omega_{t,n,p}$. This led to a design-independent upper bound on $\text{tr}(C_d)$ which is attained by d^*. Then the theorem follows by invoking Theorem 1.4.1.

Remark 7.3.1. We may recall from Chapter 4 that under model (1.3.14), Kushner (1997b, 1999) gave designs which are optimum in the entire class $\Omega_{t,n,p}$ and these were found to be non-binary. However, in experimental situations, binary designs are often preferred by practitioners and in that context, Theorem 7.3.1 is useful. Moreover, these designs may be constructed with $n = t(t-1)$ which could sometimes be smaller than the number of subjects of the exact optimal designs obtained via approximate theory as in Chapter 4. The efficiency of the designs d^* as given by Theorem 7.3.1 in the general class $\Omega_{t,n,p}$ is quite high and for more discussion on this we refer to Kunert and Martin (2000b). □

Hedayat and Yan (2008) considered model (5.2.1), i.e., a model with self and mixed carryover effects, but instead of assuming uncorrelated errors as in Chapter 5, they adopted an error dispersion matrix of the form $\sigma^2\Sigma = \sigma^2 I_n \otimes V$ as used with model (1.3.14). Then again (7.3.1) holds, with C_d and $\tilde{C}_d$ now defined with reference to this model. Moreover, if a design $d^* \in \Omega_{t,n,p}$ is given by an $OA_I(n, p, t, 2)$, with the rows and columns of the array representing the periods and subjects of d^*, they proved that even under this model, C_{d^*} is completely symmetric and $C_{d^*} = \tilde{C}_{d^*}$. For maximizing $\text{tr}(\tilde{C}_d)$, Hedayat and Yan (2008) considered two particular forms of V, one where V is as given in (7.2.4) and another where the errors within each subject follow a stationary first order moving average process. Under the latter form,

$$V = I_p + \rho W, \quad \rho \in (-1/2, 1/2) \tag{7.3.3}$$

where W is a $p \times p$ matrix with 1 in the cells $(i, i+1)$ and $(i+1, i)$, and 0 in all other cells. Thus errors from adjacent periods for the same subject are correlated, while all other errors are uncorrelated. For such a model they gave the following result.

Theorem 7.3.2. *For $p = 3$ and $t \geq 3$, let $d^* \in \Omega_{t,n,3}$ be a design given by an orthogonal array of type I, $OA_I(n, 3, t, 2)$. Then d^* is universally optimal for the estimation of direct effects over $\Omega_{t,n,3}$ under a model with self and mixed carryovers with the dispersion structure of errors within each subject given by either*
(i) V as in (7.2.4) and any $\rho \in (-1, 1)$, or
(ii) V as in (7.3.3) and any $\rho \in (-1/2, 1/2)$. □

The above theorem follows on showing that d^* maximizes $\text{tr}(\tilde{C}_d)$ over the entire class $\Omega_{t,n,3}$. The proof uses the techniques of Kunert and Martin (2000a) which were also used to prove the results of Section 5.2. Although this optimality result was established for $p = 3$, in contrast to Theorem 7.3.1, no restriction to the binary class was required for this purpose. Hedayat and Yan (2008) gave some numerical results for $p = 4$ and $t = 4, 5$ and 7 which show that the design d^* obtained from an $OA_I(n, 4, t, 2)$ is highly efficient. We give two examples below.

Example 7.3.2. (a) When $p = 3$, the design given by the array $OA_I(6, 3, 3, 2)$ shown in Example 3.1.1 is universally optimal for the estimation of direct effects over $\Omega_{3,6,3}$ under a model with self and mixed carryovers and with a correlated error structure given by either of the two forms (7.2.4) or (7.3.3).
(b) When $p = 4$, Hedayat and Yan (2008) observed from numerical computations that the design in $\Omega_{4,12,4}$ given by the $OA_I(12, 4, 4, 2)$ in Example 3.1.1 has an efficiency of at least 0.999 for some chosen values of ρ in the intervals $(-0.99, 0.99)$ and $(-0.49, 0.49)$, for V of the form (7.2.4) and (7.3.3), respectively. This shows that this design is likely to be optimal or at least highly efficient for all ρ. □

7.4 Optimal Designs for Test-Control Comparisons

In this section we consider the problem of comparing t test treatments with a standard treatment called the control treatment. Thus there are $t + 1$ treatments in all. Let $\Omega_{t+1,n,p}$ denote the class of all crossover designs with t test treatments and one control treatment applied to n subjects over p

periods. We designate the control treatment as treatment 0 while the test treatments are $1, \ldots, t$ as before. The direct and carryover effect of the control will be denoted by τ_0 and ρ_0, respectively, while those for the test treatments will be denoted by $\tau_1, \ldots, \tau_t$ and $\rho_1, \ldots, \rho_t$ as before. The direct effect contrasts of interest are the contrasts between the test treatment and the control treatment, i.e.

$$\tau_0 - \tau_i \quad 1 \leq i \leq t. \tag{7.4.1}$$

Similarly, the carryover effect contrasts of interest are $\rho_0 - \rho_i \quad 1 \leq i \leq t$.

We consider model (1.3.3) where now s ranges over $0, 1, \ldots, t$ and $\boldsymbol{\tau}$ and $\boldsymbol{\rho}$ are the $(t+1) \times 1$ vectors

$$\boldsymbol{\tau} = (\tau_0, \tau_1, \ldots, \tau_t)' \quad \text{and} \quad \boldsymbol{\rho} = (\rho_0, \rho_1, \ldots, \rho_t)'.$$

Under this model, for a design $d \in \Omega_{t+1,n,p}$, the information matrix for direct effects is given by C_d as in (1.3.13), with C_d now having $t+1$ rows and columns. While comparing t test treatments with a single control, uniformity over periods is not desirable as all treatments are not equally important and we would like to have larger replication for the control. On the other hand, as seen before, uniformity over periods plays an important role in deriving results on optimality of crossover designs as it allows the equality to hold in inequalities of the form given by (7.3.1). Hedayat and Yang (2005) made the interesting observation that even in the present context of test-control comparisons, such an equality holds for a design d as long as d has

$$m_{dsi} = \frac{r_{ds}}{p}, \quad 0 \leq s \leq t, \quad 1 \leq i \leq p,$$

even though the replication of each test treatment and control treatment may be different. So, for such designs we can, as before, ignore period effects in a model while computing C_d.

From (7.4.1) it is clear that for direct effects, the parametric function of interest is the contrast vector

$$\begin{pmatrix} \tau_0 - \tau_1 \\ \cdots \\ \tau_0 - \tau_t \end{pmatrix} = \begin{pmatrix} 1 & -1 & 0 \cdots & 0 \\ 1 & 0 & -1 \cdots & 0 \\ \vdots & & & \\ 1 & 0 & 0 \cdots & -1 \end{pmatrix} \boldsymbol{\tau} = B\boldsymbol{\tau}, \ \text{(say)}. \tag{7.4.2}$$

Let $\mathcal{I}_d$ be the information matrix for $B\boldsymbol{\tau}$ under model (1.3.3). Then the following relation between $\mathcal{I}_d$ and C_d is easy to see.

Lemma 7.4.1. *Let $\bar{I}$ be the $(t+1) \times t$ matrix given by*

$$\bar{I} = \begin{pmatrix} \mathbf{0}'_t \\ I_t \end{pmatrix}.$$

Then for any $d \in \Omega_{t+1,n,p}$,

$$\mathcal{I}_d = \bar{I}' C_d \bar{I}, \tag{7.4.3}$$

and hence $\mathcal{I}_d$ can be obtained from C_d simply by deleting its first row and first column. □

So far in this book we have studied designs for comparing t equally important treatments and for this problem, the universal optimality criterion has generally been used. However, for the present problem of comparing t treatments against a control, (i.e., for inference on the contrast vector $B\boldsymbol{\tau}$), the A- and MV-optimality criteria seem natural and are frequently used. These are defined below.

Definition 7.4.1. A design $d^* \in \Omega_{t+1,n,p}$ is said to be A-optimal for $B\boldsymbol{\tau}$ if it minimizes $\text{tr}(\mathcal{I}_d^{-1})$ over all $d \in \Omega_{t+1,n,p}$.

Definition 7.4.2. A design $d^* \in \Omega_{t+1,n,p}$ is said to be MV-optimal for $B\boldsymbol{\tau}$ if it minimizes $\max\limits_{1 \le s \le t} \text{var}(\hat{\tau}_0 - \hat{\tau}_s)$ over all $d \in \Omega_{t+1,n,p}$, where $\hat{\tau}_0 - \hat{\tau}_s$ is the best linear unbiased estimator (BLUE) of $\tau_0 - \tau_s$, $1 \le s \le t$.

Thus, while the A-optimality criterion aims at minimizing the sum of variances of BLUE's of the contrasts of interest, the MV-optimality criterion calls for minimizing the maximum of these variances. Similar definitions apply for the A- and MV-optimality for the carryover effects for comparing test treatments with a control treatment. The following connection between the above two criteria is well known.

Lemma 7.4.2 *An A-optimal design d is also MV-optimal if $\mathcal{I}_d$ is completely symmetric.* □

If t, p and n are such that an optimal design cannot be found, one would like to use a design with high A-efficiency. As usual, the A-efficiency of any design $d \in \Omega_{t+1,n,p}$ is given by

$$\text{tr}(\mathcal{I}_{d^*}^{-1})/\text{tr}(\mathcal{I}_d^{-1}) \times 100\%,$$

where d^* is the A-optimal design in $\Omega_{t+1,n,p}$. On replacing $\text{tr}(\mathcal{I}_{d^*}^{-1})$ by its lower bound in the class, one can obtain a lower bound for the A-efficiency.

The problem of finding suitable crossover designs for making test versus control treatment comparisons was first considered by Pigeon and

Raghavarao (1987). They defined a class of designs called *control balanced residual effects designs*, gave methods of construction of these designs but did not investigate their optimality properties. Some of these designs were later proved to be optimal by Majumdar (1988) who initiated the study of optimal crossover designs for test-control comparisons. He showed that given a strongly balanced uniform design with the t test treatments, some of the treatment labels in it can be suitably replaced by the control to obtain an optimal design. However, since a strongly balanced uniform design was used as the starting point, these optimal designs have a large number of periods. Subsequently, Hedayat and Yang (2005, 2006) and others studied the practically important case of $p \leq t+1$. In this section, we present a selection of these results without proof and give several illustrative examples.

7.4.1 *Designs with* $p > t + 1$

Majumdar (1988) assumed model (1.3.3) and obtained A- and MV-optimal designs for $B\boldsymbol{\tau}$. Starting with an optimal design under a simpler version of model (1.3.3) where all effects other than the direct effect are excluded, he ensured that appropriate orthogonality conditions are met so as to render this design optimal under the finer model (1.3.3) too. He observed that a design $d^* \in \Omega_{t+1,n,p}$ is A-optimal (and also MV-optimal) for the test control comparisons of direct effects under the simpler model if

$$r_{d^*1} = \cdots = r_{d^*t}, \quad r_{d^*0} = r_{d^*1}t^{\frac{1}{2}}, \tag{7.4.4}$$

and the relevant orthogonality conditions are met if d^* satisfies

$$\begin{aligned} m_{d^*1i} &= \cdots = m_{d^*ti}, \quad m_{d^*0i} = m_{d^*1i}t^{\frac{1}{2}}, \ 1 \leq i \leq p, \\ n_{d^*1j} &= \cdots = n_{d^*tj}, \quad n_{d^*0j} = n_{d^*1j}t^{\frac{1}{2}}, \ 1 \leq j \leq n, \end{aligned} \tag{7.4.5}$$

and

$$\begin{aligned} z_{d^*10} &= \cdots = z_{d^*t0}, \quad z_{d^*00} = z_{d^*10}t^{\frac{1}{2}}, \\ z_{d^*1s} &= \cdots = z_{d^*ts}, \quad z_{d^*0s} = z_{d^*1s}t^{\frac{1}{2}}, \ 1 \leq s \leq t, \end{aligned} \tag{7.4.6}$$

with the notation in (7.4.4)–(7.4.6) being as defined in Chapter 1. Thus any design satisfying (7.4.4)–(7.4.6) is A-optimal (and also MV-optimal) for the test control comparisons of direct effects under model (1.3.3). Majumdar (1988) identified designs satisfying (7.4.4)–(7.4.6) and hence obtained the following result.

Theorem 7.4.1. *Let $t = c^2$ for a positive integer c and let $d_0 \in \Omega_{c^2+c,n,p}$ be a strongly balanced uniform crossover design. Let d^* be a design obtained from d_0 by replacing each of the treatment labels $c^2+1, c^2+2, \ldots, c^2+c$ by* *the control treatment label* 0*, keeping everything else unchanged. Then* d^* *is* A*- and* MV*-optimal for direct effects for comparing the* c^2 *test treatments with the control over* $\Omega_{c^2+1,n,p}$. □

Example 7.4.1. Let $t = 4$. Here $c = 2$ and a strongly balanced uniform design in $\Omega_{6,36,12}$ can be constructed using the method given in Theorem 2.6.2. Now, following Theorem 7.4.1 and replacing treatment labels 5 and 6 in this design by the control treatment label 0, the optimal design $d^* \in \Omega_{4+1,36,12}$ for test-control comparisons may be obtained as shown below.

111111	222222	333333	444444	000000	000000
123400	123400	123400	123400	123400	123400
222222	333333	444444	000000	000000	111111
234001	234001	234001	234001	234001	234001
333333	444444	000000	000000	111111	222222
340012	340012	340012	340012	340012	340012
444444	000000	000000	111111	222222	333333.
400123	400123	400123	400123	400123	400123
000000	000000	111111	222222	333333	444444
001234	001234	001234	001234	001234	001234
000000	111111	222222	333333	444444	000000
012340	012340	012340	012340	012340	012340

□

Ting (2002) employed tools similar to those used by Majumdar (1988) and obtained some additional optimal crossover designs for comparing t test treatments with a control for some specific combinations of t, n and p with $p > t + 1$. Example 7.4.2 below gives an A-optimal design given by Ting (2002) for $t = 4$. Note that this has fewer subjects than the design in Example 7.4.1 but requires one more period.

In cases where an A-optimal design could not be found, Ting (2002) gave A-efficient designs which can be obtained from existing universally optimal designs in test treatments by replacing some of its treatment labels by 0. In Example 7.4.3, we give two such efficient designs for control-test comparisons.

Example 7.4.2. Let $t = 4, n = 6, p = 13$. The following design is A-optimal in $\Omega_{4+1,6,13}$:

$$\begin{array}{cccccc} 0&0&1&4&3&2\\ 0&0&2&3&4&1\\ 3&1&0&2&0&4\\ 1&3&0&2&0&4\\ 1&3&2&0&4&0\\ 0&0&4&1&2&3\\ 4&2&1&0&3&0\\ 3&1&0&4&0&2\\ 4&2&3&0&1&0\\ 2&4&3&0&1&0\\ 2&4&0&1&0&3\\ 0&0&4&3&2&1\\ 0&0&1&2&3&4 \end{array}.$$

□

Example 7.4.3. Let $t = 2$. Consider the two designs $d_1 \in \Omega_{3,12,5}$ and $d_2 \in \Omega_{3,9,6}$ given in Examples 2.5.1 and 2.2.1, respectively. Both of these are universally optimal for direct effects (see Theorems 2.4.1 and 2.5.3). Replacing 3 by 0 in d_1 gives a design $d_1^* \in \Omega_{2+1,12,5}$ with A-efficiency equal to 0.9325. Again, replacing treatment label 3 by the control label 0 in d_2, one obtains a design $d_2^* \in \Omega_{2+1,9,6}$ which has A-efficiency equal to 0.9714. These efficient designs are shown below.

$$d_1^* = \begin{array}{c} 2\,1\,1\,0\,1\,0\,1\,2\,0\,2\,2\,0\\ 1\,2\,1\,2\,0\,1\,1\,0\,2\,0\,2\,0\\ 1\,2\,0\,2\,0\,1\,2\,0\,2\,0\,1\,1\\ 2\,1\,2\,1\,1\,0\,0\,1\,0\,2\,0\,2\\ 0\,0\,2\,1\,2\,2\,0\,1\,1\,1\,0\,2 \end{array}, \quad d_2^* = \begin{array}{c} 1\,1\,1\,2\,2\,2\,0\,0\,0\\ 1\,2\,0\,1\,2\,0\,1\,2\,0\\ 2\,2\,2\,0\,0\,0\,1\,1\,1\\ 2\,0\,1\,2\,0\,1\,2\,0\,1\\ 0\,0\,0\,1\,1\,1\,2\,2\,2\\ 0\,1\,2\,0\,1\,2\,0\,1\,2 \end{array}.$$

□

The A- and MV-optimal designs of Theorem 7.4.1 require a large number of periods and subjects since $p = \mu_2(t+\sqrt{t})$ and $n = \mu_3(t+\sqrt{t})^2$, where $\mu_3 \geq 1, \mu_2 \geq 2$. The smallest such design exists for $t = 4(= 2^2)$ and with $\mu_3 = 1$ and $\mu_2 = 2$, this design is shown in Example 7.4.1. It requires 12 periods. The A-optimal (or, highly efficient) designs of Ting (2002) also have $p > t + 1$.

Since designs with a large number of periods are often unattractive to experimenters, it is important to derive optimal designs where the number of periods is less than the number of treatments, and so, in the present context, it is desirable that $p \leq t+1$. In the following two subsections, some of the available optimality results for the case $p \leq t+1$ are reviewed.

7.4.2 *Designs with* $p = 2$

We have discussed the optimality aspects of two-period crossover designs in Section 3.3 wherein all treatments were considered equally important. In this subsection, we state some results on optimal two-period designs for test-control treatment comparisons. Using the available results on the A- and MV-optimality of block designs for comparing test treatments with a control and the connection established between certain block designs and crossover designs (see Section 3.3), Hedayat and Zhao (1990) obtained the following results. Here too, as in Theorem 7.4.1, the optimal design is obtained by starting from another design with a larger number of treatments and then replacing some treatment labels in the latter by the control treatment label 0.

Theorem 7.4.2. *Let* $t = c^2$ *for a positive integer* c, $n \equiv 0 \pmod{t}$ *and* $d_0 \in \Omega_{c^2+c,n,2}$ *be the design as given by Theorem 3.3.3. Let* d^* *be a design obtained from* d_0 *by replacing each of the treatment labels* $c^2+1, c^2+2, \ldots, c^2+c$ *by the control treatment label* 0, *keeping everything else unchanged. Then* d^* *is* A- *and* MV*-optimal for direct effects for comparing* c^2 *test treatments with a control over* $\Omega_{c^2+1,n,2}$. □

The conclusion of Theorem 7.4.2 remains valid for carryover effects when $d_0 \in \Omega_{c^2+c,n,2}$ is taken as the design given by Theorem 3.3.5. Furthermore, as $p = 2$, the optimality result in Theorem 7.4.2, or its counterpart for carryover effects, holds not only under model (1.3.3) but also under model (1.3.14) with correlated error structure; see Lemma 1.3.1.

Example 7.4.4. Let $t = 4$. We first need to construct an optimal design d_0 with $t = 6(= 4 + \sqrt{4})$ as in Theorem 3.3.3. Suppose we take $n = 18$ and $f_{d_0,s} = 6$ for $s = 1,2,3$ and $f_{d_0,s} = 0$ for $s = 4,5,6$. Then, by replacing treatment labels 5 and 6 by 0 in d_0, we get the following design. By Theorem 7.4.2, this design is A-optimal for direct effects for comparing

4 test treatments with a control over $\Omega_{4+1,18,2}$.

$$\begin{matrix} 1&1&1&1&1&1&2&2&2&2&2&2&3&3&3&3&3&3 \\ 1&2&3&4&0&0&1&2&3&4&0&0&1&2&3&4&0&0 \end{matrix}.$$

□

When t does not divide n, Hedayat and Zhao (1990) obtained an A-optimal design similar to the design in Theorem 3.3.4 where the first period has only one treatment. Their result is stated below.

Theorem 7.4.3. *Suppose t does not divide n and let $d^* \in \Omega_{t+1,n,2}$ be a design where only one treatment (either the control or a test treatment) is assigned to the first period. In the second period, all test treatments are applied as equally often as possible and the control treatment is applied r_0 times, where r_0 is a positive integer which minimizes*

$$\frac{t}{r_0} + \frac{t - n + r_0 + tx}{t(r_0 + (n - r_0 - tx)/(x + 1))}$$

with $x = \left[\dfrac{n - r_0}{t}\right]$ and as before, $[z]$ denotes the largest integer not exceeding $z > 0$. Then d^* is A-optimal for direct effects for comparing t test treatments with a control over $\Omega_{t+1,n,2}$. □

For some other results on two-period crossover designs for comparing *two* test treatments with a control, see Koch, Amara, Brown, Colton and Gillings (1989).

7.4.3 *Designs with* $3 \leq p \leq t + 1$

Several interesting results on optimal crossover designs for making test-control comparisons have been obtained by Hedayat and Yang (2005, 2006), Yang and Park (2007), Hedayat and Yan (2008) and Yang and Stufken (2008). Some of these results are highlighted in this section, with notation as defined in Chapter 1.

Hedayat and Yang (2005) considered a subclass of designs for which
(a) the control treatment appears equally often in the p periods, and
(b) no treatment precedes itself.
They gave a characterization of designs which are both A-optimal and MV-optimal over this subclass. Denoting this subclass by $\Lambda_{t+1,n,p}$, we may write

$$\begin{aligned} \Lambda_{t+1,n,p} = \{d : d \in \Omega_{t+1,n,p},\ (a)\ m_{d0i} = r_{d0}/p \ \text{ for } \ 1 \leq i \leq p, \\ \text{and } \ (b)\ z_{dss} = 0, \ \text{ for } \ 0 \leq s \leq t\}. \end{aligned} \tag{7.4.7}$$

After some lengthy and elaborate algebra, Hedayat and Yang (2005) gave a bound for $\mathrm{tr}(\mathcal{I}_{\mathrm{d}}^{-1})$ and defined a class of designs in $\Lambda_{t+1,n,p}$ for which this bound is attained. We state their bound in the following lemma and define this class of designs below.

Lemma 7.4.3. *For $3 \le p \le t+1$, and all $d \in \Lambda_{t+1,n,p}$ with r_{d0} fixed,*

$$\mathrm{tr}(\mathcal{I}_d^{-1}) \ge \frac{t(t-1)^2 p}{x} + \frac{tp}{y} \tag{7.4.8}$$

where

$$x = t(p-1)(np - r_{d0}) - p(r_{d0} - p^{-1} \min \sum_{j=1}^{n} n_{d0j}^2) - \frac{[nt(p-1) - t\bar{r}_{d0} - \min \sum_{j=1}^{n} n_{d0j}\bar{n}_{d0j}]^2}{n(p-1)(pt-t-1) - (pt-t+p-2)\bar{r}_{d0} + \min \sum_{j=1}^{n} \bar{n}_{d0j}^2}$$

and

$$y = p(r_{d0} - p^{-1} \min \sum_{j=1}^{n} n_{d0j}^2) - \frac{n(p-1)(\min \sum_{j=1}^{n} n_{d0j}\bar{n}_{d0j})^2}{np(p-1)\bar{r}_{d0} - \bar{r}_{d0}^2 - n(p-1)\sum_{j=1}^{n} \bar{n}_{d0j}^2}.$$

Equality holds in (7.4.8) when d satisfies the following conditions.
(i) $m_{dsi} = r_{ds}/p$, $0 \le s \le t$,
(ii) $T_d'\mathrm{pr}^{\perp}(U)T_d$, $T_d'\mathrm{pr}^{\perp}(U)F_d$, $F_d'\mathrm{pr}^{\perp}(U)F_d$ are invariant under any permutation of test treatments,
(iii) each test treatment appears at most once in each subject in the first $p-1$ periods,
(iv) $n_{dsj} = 0$ or 1, $1 \le s \le t$; $1 \le j \le n$, and
(v) $\sum_{j=1}^{n} n_{d0j}^2$, $\sum_{j=1}^{n} n_{d0j}\bar{n}_{d0j}$ and $\sum_{j=1}^{n} \bar{n}_{d0j}^2$ are minimized over all designs in $\Lambda_{t+1,n,p}$ with r_{d0} fixed. □

Following Hedayat and Yang (2005), we now give the definitions of some designs in $\Omega_{t+1,n,p}$ with $3 \le p \le t+1$. As usual, the subjects are taken as blocks. These definitions are analogous to Definitions 5.2.1–5.2.3.

Definition 7.4.3. A design $d \in \Omega_{t+1,n,p}$ is called a balanced test-control incomplete block design for the direct effects if
(a) each of the t test treatments appears equally often in d,
(b) in each subject, each test treatment appears at most once,
(c) the number of subjects where test treatments s and s' appear together is the same for every pair s, s', $s \ne s', 1 \le s, s' \le t$,
(d) the control treatment appears in each subject either $[r_{d0}/n]$ or $[r_{d0}/n]+1$ times, and

(e) the number of subjects where the control treatment appears $[r_{d0}/n]$ times and the test treatment s appears is the same for every s, $1 \leq s \leq t$.

Definition 7.4.4. A design $d \in \Omega_{t+1,n,p}$ is called a balanced test-control incomplete block design for the carryover effects if the first $p-1$ periods of the design form a balanced test-control incomplete block design for the direct effects in $\Omega_{t+1,n,p-1}$.

Definition 7.4.5. A design $d \in \Omega_{t+1,n,p}$ is called balanced for test-control carryover effects if
(a) every test treatment is immediately preceded by every other test treatment equally often,
(b) the control treatment is immediately preceded by every test treatment equally often,
(c) every test treatment is immediately preceded by the control treatment equally often, and
(d) no treatment, including the control, is immediately preceded by itself.

Definition 7.4.6. A design $d \in \Omega_{t+1,n,p}$ is called a proportional frequency design for test-control on the periods, if every test treatment appears in each period $(np - r_{d0})/(tp)$ times and the control treatment appears in each period r_{d0}/p times.

Definition 7.4.7. A design $d \in \Omega_{t+1,n,p}$ is called a totally balanced test-control incomplete crossover design if
(a) d is a balanced test-control incomplete block design for the direct effects,
(b) d is a balanced test-control incomplete block design for the carryover effects,
(c) d is balanced for test-control carryover effects,
(d) d is a proportional frequency design for test-control on the periods,
(e) the number of subjects where test treatment s appears once in the first $p-1$ periods and test treatment s' appears once is the same for every $s, s', s \neq s'; 1 \leq s, s' \leq t$,
(f) the number of subjects where the control treatment appears $[\tilde{r}_{d0}/n]$ times in the first $p-1$ periods and test treatment s appears once is the same for every test treatment, $1 \leq s \leq t$,
(g) the number of subjects where test treatment s appears once in the first $p-1$ periods and the control treatment appears $[r_{d0}/n]$ times is the same for every test treatment, $1 \leq s \leq t$.

It can easily be seen from the above definitions that conditions (i)–(v) in Lemma 7.4.3 are satisfied by a totally balanced test-control incomplete

crossover design and so, for these designs, equality holds in (7.4.8). Now, using arguments which take into account the facts that
(a) the minima of the quantities in condition (v) of Lemma 7.4.3 are functions of r_{d0}, and
(b) $\mathcal{I}_d$ is completely symmetric in view of condition (ii) of this lemma,
Hedayat and Yang (2005) obtained the following result.

Theorem 7.4.4. *For $p \leq t+1$, let a design $d^* \in \Lambda_{t+1,n,p}$ be such that*
(i) d^* is a totally balanced test-control incomplete crossover design, and
(ii) r_{d^*0} minimizes the expression on the right of (7.4.8).
Then d^* is A- and MV-optimal for direct effect for comparing t test treatments with a control treatment over $\Lambda_{t+1,n,p}$. □

It is not easy to construct designs satisfying the conditions of Theorem 7.4.4. Proposition A.6 in Hedayat and Yang (2005) gives expressions for the minima of the quantities in condition (v) of Lemma 7.4.3 as functions of r_{d0}. Using these expressions, one can use a computer program to numerically find a value of r_{d^*0} which will satisfy condition (ii) of Theorem 7.4.4. Then with this r_{d^*0}, one has to construct a design d^* satisfying condition (i) of this theorem. Hedayat and Yang (2005) stated that it is difficult to give a general method for constructing designs for which $r_{d^*0} \neq n$. If $r_{d^*0} = n$, then the construction is relatively easy and for this case, the authors gave construction procedures for the optimal designs of Theorem 7.4.4. Their construction utilizes the following results.

Lemma 7.4.4. *For given t, p, n and $r_{d0} = n$, a totally balanced test-control incomplete crossover design exists only if both $\dfrac{n(p-1)}{pt}$ and $\dfrac{(p-1)(p-2)n}{pt(t-1)}$ are integers.*

Proof. Follows from Definition 7.4.5. □

Lemma 7.4.5. *For given t, $p = t+1$, n and $r_{d0} = n$, a totally balanced test-control incomplete crossover design exists if there exists a balanced uniform design in $\Omega_{t+1,n,p}$.*

Proof. If in the balanced uniform design, treatment $t+1$ is replaced by the control treatment 0, then the resulting design is the required totally balanced test-control crossover design. □

Lemma 7.4.6. *For given $t, p = t, n$ and $r_{d0} = n$, a totally balanced test-control incomplete crossover design exists if*

(i) n/t^2 is an integer and t is a composite number, or
(ii) n/t^2 is an even integer and t is a prime number.

Proof. By the conditions (i) and (ii) of the lemma, n/t is a multiple of t when t is a composite number; moreover, when t is a prime, n/t is an even multiple of t. Therefore, as discussed in Section 2.6, there always exists a balanced uniform design in $\Omega_{t,n/t,t}$. Now the required totally balanced test-control crossover design can be constructed as follows:
(i) construct a balanced uniform design $\bar{d}$ in $\Omega_{t,n/t,t}$;
(ii) replace label 1 by label 0 in $\bar{d}$ to obtain a new design;
(iii) repeat step (ii) above for each of the labels $2, \ldots, t$ to obtain $t - 1$ additional new designs;
(iv) juxtapose the t new designs obtained in steps (ii) and (iii) to form one design with n subjects.
This is the required design. □

Lemmas 7.4.5 and 7.4.6 relate to the construction of optimal designs with $r_{d0} = n$ when $p = t + 1$ and $p = t$. For obtaining designs with $p \leq t$, and in particular with $p < t$, Hedayat and Yang (2005) suggested an approach involving the following steps:

Approach 7.4.1. (i) Construct a balanced incomplete block design (BIBD) involving t test treatments and b blocks of size $p - 1$.

(ii) Construct p arrays, each with p rows and b columns, as follows. In the ith array, 0 is placed in all b positions of the ith row, $1 \leq i \leq p$. The $p - 1$ positions thus left empty in each of the b columns of this array are filled arbitrarily by the b blocks of the BIBD in step (i). This gives the ith array, $1 \leq i \leq p$. These p arrays are juxtaposed to produce a $p \times pb$ array.

(iii) Rearrange the positions of the test treatments in each column of the $p \times pb$ array obtained in (ii) to convert this array into a totally balanced test-control incomplete crossover design.

For $p < t$, Approach 7.4.1 does not always ensure that it will lead to a totally balanced test-control incomplete crossover design. However, as Hedayat and Yang (2005) pointed out, these steps are indeed useful in constructing the desired designs when they exist. In cases where $r_{d0} \neq n$, they suggested that one can compare the minimum value on the right of (7.4.8) with its value when $r_{d0} = n$, and if, these two values are very close, then one can use Lemmas 7.4.5 and 7.4.6 to obtain A-efficient designs.

Focusing on the cases where $p = 3, 4$ or 5, Hedayat and Yang (2005) gave several examples of totally balanced test-control incomplete crossover designs with $r_{d^*0} = n$, each of these being both A- and MV-optimal for direct effects for comparing t test treatments with a control. They observed that using the methods of Lemma 7.4.5 or 7.4.6 or Approach 7.4.1, such optimal designs could be constructed for the combinations of t, n and p as in Table 7.4.1; and in each of these cases, r_{d^*0} was found to be equal to n.

TABLE 7.4.1
Values of t, n and p

p	3	3	3	3	4	4	4	4	4	5
t	2	3	4	5	3	4	6	7	9	4
n	6	9	36	30	4	16	40	28	48	10

These optimal designs (except for the ones with $p = 3, t = 2$ and $p = 3, t = 4$) are for the minimum value of n, as given by Lemma 7.4.4, for the corresponding values of t and p. Some examples of these designs are given in Examples 7.4.5–7.4.10; the first two examples being constructed by the methods of Lemmas 7.4.5 and 7.4.6, respectively, while the others were obtained by Hedayat and Yang (2005) using Approach 7.4.1.

Example 7.4.5. $p = 5, t = 4$. By Lemma 7.4.4, we may use $n = 10$ and here $r_{d^*0} = 10$. A balanced uniform design with $t = p = 5, n = 10$ can be constructed as in Theorem 2.6.1. Then as in Lemma 7.4.5, the following design is obtained from it by replacing 5 by 0.

$$\begin{array}{ccccc ccccc}
1&2&3&4&0&3&4&0&1&2\\
0&1&2&3&4&4&0&1&2&3\\
2&3&4&0&1&2&3&4&0&1\\
4&0&1&2&3&0&1&2&3&4\\
3&4&0&1&2&1&2&3&4&0
\end{array}.$$

Similarly, an optimal design with $p = 4, t = 3, n = 4$ may be constructed. □

Example 7.4.6. Let $p = 4, t = 4, n = 16$. Here $r_{d^*0} = 16$. As in Lemma 7.4.6, the following design is obtained based on the design d_1 in Example 2.2.1.

$$\begin{array}{cccc} 0\,2\,3\,4 & 1\,0\,3\,4 & 1\,2\,0\,4 & 1\,2\,3\,0 \\ 4\,0\,2\,3 & 4\,1\,0\,3 & 4\,1\,2\,0 & 0\,1\,2\,3 \\ 2\,3\,4\,0 & 0\,3\,4\,1 & 2\,0\,4\,1 & 2\,3\,0\,1 \\ 3\,4\,0\,2 & 3\,4\,1\,0 & 0\,4\,1\,2 & 3\,0\,1\,2 \end{array}.$$

□

Example 7.4.7. $p = 3, t = 3$. Here, we use $n = 9$ and again, $r_{d^*0} = 9$. Starting from the BIB design with $t = 3$ and three blocks of size two each, Approach 7.4.1 gives the following design.

$$\begin{array}{ccc} 0\,0\,0 & 2\,3\,1 & 2\,3\,1 \\ 1\,2\,3 & 0\,0\,0 & 1\,2\,3 \\ 2\,3\,1 & 1\,2\,3 & 0\,0\,0 \end{array}.$$

Similarly, an optimal design with $t = 4, p = 4, n = 16$ may be constructed.

□

Example 7.4.8. $p = 3, t = 5$. From Lemma 7.4.4, we may use $n = 30$. Starting from a BIB design with $t = 5$ and ten blocks of size two each, using Approach 7.4.1 we can construct the following design.

$$\begin{array}{ccc} 0\,0\,0\,0\,0\,0\,0\,0\,0\,0 & 1\,1\,4\,5\,3\,2\,2\,4\,3\,5 & 1\,1\,4\,5\,3\,2\,2\,4\,3\,5 \\ 2\,3\,1\,1\,2\,4\,5\,3\,5\,4 & 0\,0\,0\,0\,0\,0\,0\,0\,0\,0 & 2\,3\,1\,1\,2\,4\,5\,3\,5\,4 \\ 1\,1\,4\,5\,3\,2\,2\,4\,3\,5 & 2\,3\,1\,1\,2\,4\,5\,3\,5\,4 & 0\,0\,0\,0\,0\,0\,0\,0\,0\,0 \end{array}.$$

□

Example 7.4.9. $p = 4, t = 7, n = 28$. The following design is constructed using Approach 7.4.1, starting from a BIB design with $t = 7$ and seven blocks of size three each.

$$\begin{array}{cccc} 0\,0\,0\,0\,0\,0\,0 & 3\,4\,5\,6\,7\,1\,2 & 1\,2\,3\,4\,5\,6\,7 & 4\,5\,6\,7\,1\,2\,3 \\ 1\,2\,3\,4\,5\,6\,7 & 0\,0\,0\,0\,0\,0\,0 & 4\,5\,6\,7\,1\,2\,3 & 3\,4\,5\,6\,7\,1\,2 \\ 3\,4\,5\,6\,7\,1\,2 & 4\,5\,6\,7\,1\,2\,3 & 0\,0\,0\,0\,0\,0\,0 & 1\,2\,3\,4\,5\,6\,7 \\ 4\,5\,6\,7\,1\,2\,3 & 1\,2\,3\,4\,5\,6\,7 & 3\,4\,5\,6\,7\,1\,2 & 0\,0\,0\,0\,0\,0\,0 \end{array}.$$

□

When $r_{d^*0} \neq n$, the construction of optimal designs becomes extremely hard and so, a systematic construction method is not available. Hedayat and Yang (2005) gave two examples of such designs, one in $\Omega_{7+1,49,3}$ with

$r_{d^*0} = 42$ and the other in $\Omega_{4+1,48,5}$ with $r_{d^*0} = 60$. We exhibit one of these designs below.

Example 7.4.10. $p = 3, t = 7, n = 49, r_{d^*0} \neq n$.

$$\begin{array}{ccccccc} 0000000 & 0000000 & 1234567 & 4567123 & 4567123 & 1234567 & 4567123 \\ 3456712 & 1234567 & 0000000 & 0000000 & 1234567 & 3456712 & 3456712 \\ 4567123 & 4567123 & 3456712 & 3456712 & 0000000 & 0000000 & 1234567 \end{array}.$$

□

For some values of t, n, p, when $r_{d^*0} \neq n$, Hedayat and Yang (2005) gave A-efficient designs. The following example illustrates this. It remains to be investigated whether such designs are indeed optimal.

Example 7.4.11. Let $p = 5, t = 6$. Then, by Lemma 7.4.4, an optimal design d^* as in Theorem 7.4.4 will exist only if n is divisible by 75. Such a large n may not be feasible in some experimental situations. Instead, with $n = 30$ it was found that $r_{d^*0} = 30$. Consider the following design with $n = 30$.

$$\begin{array}{ccccc} 0\,0\,0\,0\,0\,0 & 6\,1\,2\,3\,4\,5 & 2\,3\,4\,5\,6\,1 & 3\,4\,5\,6\,1\,2 & 5\,6\,1\,2\,3\,4 \\ 1\,2\,3\,4\,5\,6 & 0\,0\,0\,0\,0\,0 & 6\,1\,2\,3\,4\,5 & 5\,6\,1\,2\,3\,4 & 6\,1\,2\,3\,4\,5 \\ 6\,1\,2\,3\,4\,5 & 2\,3\,4\,5\,6\,1 & 0\,0\,0\,0\,0\,0 & 2\,3\,4\,5\,6\,1 & 2\,3\,4\,5\,6\,1 \\ 2\,3\,4\,5\,6\,1 & 5\,6\,1\,2\,3\,4 & 3\,4\,5\,6\,1\,2 & 0\,0\,0\,0\,0\,0 & 1\,2\,3\,4\,5\,6 \\ 5\,6\,1\,2\,3\,4 & 3\,4\,5\,6\,1\,2 & 4\,5\,6\,1\,2\,3 & 6\,1\,2\,3\,4\,5 & 0\,0\,0\,0\,0\,0 \end{array}.$$

This design has an A-efficiency of 99.3 % over $\Lambda_{6+1,30,5}$ and so could be recommended in place of the optimal design. □

For $p = 5$ and $t = 7$, the minimum permissible value of n by Lemma 7.4.4 is $n = 35$. Hedayat and Yang (2005) observed that with $n = 70$, r_{d^*0} is also equal to 70 and they constructed a design with $n = 70$ which is a juxtaposition of two designs with $n = 35$ each. The design with $n = 70$ is optimal while each of the two component designs with $n = 35$ is highly efficient, their A-efficiencies being as high as 99.39 % and 99.36 %, respectively.

It may be recalled that the above optimality results are valid in the class $\Lambda_{t+1,n,p}$ given by (7.4.7). It is not known if the designs which are optimal in this class remain so in $\Omega_{t+1,n,p}$, though Hedayat and Yang (2005) conjectured that these designs are highly efficient, if not optimal, in the entire class. Also, they demonstrated the efficiency of these optimal designs under different simpler models (e.g., a model without carryover effects or one with only direct effects) and it turns out that these optimal designs are often highly efficient under such models.

It makes sense to study whether these optimal designs remain so in broader classes of competing designs where one or both of the two conditions (a) and (b) on designs in $\Lambda_{t+1,n,p}$ (see (7.4.7)) are removed. This problem in general, is hard to solve, though some results partially answering this question have been obtained by Hedayat and Yang (2006) and by Yang and Park (2007) for $p = 3$. To motivate the discussion, consider the following design d_1, given by Hedayat and Yang (2006).

Example 7.4.12. $t = 4, p = 4, n = 12$.

$$d_1 = \begin{matrix} 0 & 0 & 0 & 0 & 3 & 1 & 4 & 2 & 4 & 3 & 2 & 1 \\ 2 & 4 & 1 & 3 & 0 & 0 & 0 & 0 & 3 & 1 & 4 & 2 \\ 3 & 1 & 4 & 2 & 0 & 0 & 0 & 0 & 2 & 4 & 1 & 3 \\ 4 & 3 & 2 & 1 & 2 & 4 & 1 & 3 & 0 & 0 & 0 & 0 \end{matrix}.$$

□

Note that though (a) in (7.4.7) holds, $z_{d_1 00} \neq 0$ and so, by (7.4.7), $d_1 \notin \Lambda_{4+1,12,4}$. Also, it can be seen that $\text{tr}(\mathcal{I}_{d_1}^{-1}) = 0.8733$. On the other hand, using a computer program, Hedayat and Yang (2005) showed that $\text{tr}(\mathcal{I}_d^{-1}) \geq 0.8879$ for any design $d \in \Lambda_{4+1,12,4}$. Thus this design, which is outside $\Lambda_{4+1,12,4}$, is better as per the A-criterion than any design in this class.

Hedayat and Yang (2006) relaxed condition (b) in (7.4.7) and considered the following wider class of designs:

$$\Omega^1_{t+1,n,p} = \{d : d \in \Omega_{t+1,n,p}, m_{d0i} = r_{d0}/p, \ 1 \leq i \leq p\}.$$

For $p \geq 4$, $n \geq p(p-1)/2$, and t satisfying $(p-3)(p-2)+2 \leq t \leq (p-2)(p-1)+1$, they gave a new lower bound to $\text{tr}(\mathcal{I}_d^{-1})$ for a design $d \in \Omega^1_{t+1,n,p}$. They also gave sufficient conditions on a design d^* for this bound to be attained. Though it is difficult to construct designs attaining this bound, it was used by Hedayat and Yang (2006) to obtain lower bounds for the A-efficiencies of available designs in $\Omega^1_{t+1,n,p}$. In many cases they were able to give designs that are highly efficient according to the A-criterion. The following examples illustrate this fact.

Example 7.4.13. Consider the design in Example 7.4.9 which is A- and MV-optimal in $\Lambda_{7+1,28,4}$. The A-efficiency of this design in $\Omega^1_{7+1,28,4}$ is at least 99.9 %. □

Example 7.4.14. Consider the following design with $t = 4, p = 4, n = 17$:

$$\begin{array}{ccccccccccccccccc} 0&0&0&0&3&1&4&2&4&1&2&3&4&3&2&1&0 \\ 2&4&1&3&0&0&0&0&3&2&4&1&3&1&4&0&2 \\ 3&1&4&2&4&3&2&1&0&0&0&0&2&4&0&3&1 \\ 4&3&2&1&2&4&1&3&1&4&0&0&0&0&0&2&3 \end{array}.$$

The A-efficiency of this design in $\Omega^1_{4+1,17,4}$ is at least 97.9 %. □

The simpler case of designs with only three periods was considered by Yang and Park (2007) who gave a lower bound to $\text{tr}(\mathcal{I}_d^{-1})$ in $\Omega^1_{t+1,n,3}$ which is somewhat easier to compute. Using this, they showed that the A-efficiency of the design in Example 7.4.10 (which was optimal in $\Lambda_{7+1,49,3}$) is at least 99.81% in the wider class $\Omega^1_{7+1,49,3}$. They also gave optimal designs in this class and one such optimal design is shown below.

Example 7.4.15. Let $t = 3, p = 3, n = 42$. Then the following design is both A- and MV-optimal in $\Omega^1_{3+1,42,3}$:

$$\begin{array}{ccccccc} 000000 & 000000 & 000111 & 122223 & 333111 & 122223 & 333123 \\ 123111 & 122223 & 333000 & 000000 & 000223 & 311331 & 122000 \\ 123223 & 311332 & 211223 & 311331 & 122000 & 000000 & 000000 \end{array}.$$

□

Continuing with this line of work, Yang and Stufken (2008) obtained further optimal and efficient designs. They considered various models including the traditional model (1.3.3), the model (5.2.1) with self and mixed carryover effects and several simpler variants of these two models, e.g., a model with only direct effects, a model with no period effects, a model with no carryover effects, and so on. Under these models, they obtained lower bounds for $\text{tr}(\mathcal{I}_d^{-1})$ and as before, when t, n, p are such that no optimal design can be readily found, these bounds are helpful in finding the efficiency of an available design. Yang and Stufken (2008) also gave two algorithms for generating designs that have high efficiency under several models. The range of t, n and p covered by the algorithms is $p = 3, 4$ or 5, $p - 1 \leq t \leq 9$ and $n \geq t$. For given t, n and p in this range, these algorithms essentially give a design which is close to a totally balanced test-control incomplete crossover design (which is expected to be highly efficient, if not optimal, under several models). These algorithms also give A-efficiency bounds for the generated design under the above models. Using one of the algorithms, Yang and Stufken (2008) obtained the following design.

Example 7.4.16. $t = 4, p = 3, n = 20.$

$$\begin{array}{cccccccccccccccccccc} 0&2&4&1&3&0&1&3&0&2&4&0&0&1&0&0&3&4&2&4\\ 1&4&2&0&0&3&0&0&3&3&1&1&3&2&2&4&4&3&0&0\\ 4&0&0&3&4&1&2&1&2&0&0&3&4&0&1&2&0&0&3&1 \end{array}.$$

□

This design has very high A-efficiencies under almost all the above mentioned models considered by Yang and Stufken (2008). For details of the algorithms coded in SAS, their original paper may be consulted.

7.5 Optimal Designs under Subject Dropout

In some crossover experiments it may so happen that the experiment cannot be continued for the initially planned p periods for all n subjects. For example, in experiments in the area of clinical research, it is quite common that some of the subjects (patients, in this context) drop out from the study before the entire sequence of p treatments assigned to the subjects can be completed. This problem may be quite severe and as remarked by Low, Lewis and Prescott (1999), "a dropout rate of between 5% and 10% is not uncommon and, in some areas, can be as high as 25%." For examples of such experiments with dropouts and more details, we refer to Matthews (1988), Low, Lewis, McKay and Prescott (1994), Low *et al.* (1999, 2003) and Godolphin (2004).

When subjects drop out from the experiment before the trial is completed, the final "implemented" design is a truncated form of the originally "planned" design. Let these two designs be denoted by d_{imp} and d_{plan}, respectively. This truncation results in a reduced accuracy of the inference about the direct and carryover effects as obtained from d_{imp} compared to that from d_{plan}. Clearly, even if we start with an optimal d_{plan}, the design d_{imp} may not be optimal or even efficient and, in some extreme cases, may not even remain connected. Therefore, in choosing a d_{plan} for the study, the possibility of such truncation needs to be taken into consideration. In this section, we summarize a few of the available results in this direction.

Low *et al.* (1999, 2003) gave a computer intensive method for assessing the robustness of a planned design d_{plan} with $p > 2$, against random dropout of subjects. Adopting the traditional model (1.3.3), they assumed that d_{plan} assigns g subjects to each of m distinct treatment sequences and thus $n = mg$. Further, they assumed that within each period, each subject has the same probability of dropping out, independently of the other subjects, and once a subject drops out of the study, it does not return. Under this set up, let $\mathcal{D}$ be the class of all designs d_{imp} which could possibly be implemented from d_{plan}. In order to assess the efficiency of a connected design in $\mathcal{D}$, Low *et al.* (1999) used performance measures based on the A- and MV-optimality criteria, such as the average or the maximum variance of pairwise comparisons for direct and carryover effects. To keep the computations manageable, they assumed that probability of dropouts before the final period is negligible and the probability that a subject drops out in the final (i.e., pth) period is a constant, say, θ. Since the performance measures are random variables, they computed the means and standard deviations of the measures for some chosen values of θ in the interval [0,1], and used these as summary measures of the performance of a d_{plan} for a fixed value of θ. Several d_{plan} designs with $t = 4$ were compared, e.g., designs based on replicates of a single Williams square or a pair of squares, and so on. From these computations, they found that a design based on a pair of squares is more robust against dropouts than a design based on a single square, both designs having the same n. They also conjectured that in order to be robust to dropouts in the final period, a planned design should have as many distinct ordered pairs of treatments as possible in the final two periods. However, the verification of this conjecture in a general set up is computationally very hard.

This study was extended by Majumdar, Dean and Lewis (2008) who allowed subjects to drop out at random as in Low *et al.* (1999), and studied the maximum loss of efficiency for estimating direct effects with d_{imp} instead of d_{plan} and then gave bounds for this maximum loss. The d_{plan} designs considered by them are all balanced uniform designs with $p = t$, as in the next example.

Example 7.5.1. Designs d_1 and d_2 shown below consist of a pair of Williams squares while d_3 consists of a single such square, with $t = 3, 5$ and 4, respectively. Suppose d_1, d_2 and d_3 are used as three d_{plan} designs. Then if no observation is taken on the last period, Majumdar *et al.* (2008) showed that the resulting d_{imp} designs arising from d_1 and d_2 remain connected while that from d_3 becomes disconnected. Thus the two designs based on a pair of squares are more robust to dropouts in the final period than the one based on a single square. A similar observation was also made by Low *et al.* (1999).

$$d_1 = \begin{matrix} 1\,2\,0 & 2\,0\,1 \\ 0\,1\,2 & 0\,1\,2 \\ 2\,0\,1 & 1\,2\,0 \end{matrix}, \quad d_2 = \begin{matrix} 1\,2\,3\,4\,0 & 3\,4\,0\,1\,2 \\ 0\,1\,2\,3\,4 & 4\,0\,1\,2\,3 \\ 2\,3\,4\,0\,1 & 2\,3\,4\,0\,1 \\ 4\,0\,1\,2\,3 & 0\,1\,2\,3\,4 \\ 3\,4\,0\,1\,2 & 1\,2\,3\,4\,0 \end{matrix}, \quad d_3 = \begin{matrix} 0\,1\,2\,3 \\ 1\,2\,3\,0 \\ 3\,0\,1\,2 \\ 2\,3\,0\,1 \end{matrix}.$$

□

Now consider a d_{plan} which is a balanced uniform design in $\Omega_{t,n=\mu_1 t,p=t}$. Suppose all the subjects remain in the experiment for the first $t-u$ periods and then subjects start dropping out completely at random, $1 \le u \le t-1$. The design consisting of these $t-u$ periods of d_{plan} is called the minimal design and let this design be denoted by $d_{\min}$. Clearly, among all possible d_{imp} designs resulting from d_{plan} due to subject dropout, $d_{\min}$ will have the maximum loss of efficiency. Then, for a design d, with C_d denoting as usual the information matrix for direct effects, the following is true.

Lemma 7.5.1. *The information matrices for direct effects corresponding to the designs d_{plan}, d_{imp} and $d_{\min}$ satisfy the following Loewner ordering:*

$$C_{d_{\text{plan}}} \ge C_{d_{\text{imp}}} \ge C_{d_{\min}}.$$

□

Lemma 7.5.1 implies that if $d_{\min}$ is connected, so is d_{imp} and the efficiency of $d_{\min}$ may serve as a lower bound to the efficiency of d_{imp}. Thus it suffices to study $d_{\min}$ in order to assess the maximum loss of efficiency that may occur due to subject dropouts in d_{plan}.

Majumdar *et al.* (2008) proved the following result.

Lemma 7.5.2. *Let $d_{\text{plan}} \in \Omega_{t,\mu_1 t,t}$ be a balanced uniform design and suppose $u \ge 1$ and $t \ge 2u+2$. Let $\mu_1\alpha_i$, $1 \le i \le t-1$, denote the nonzero eigenvalues of $C_{d_{\min}}$. Then, for $1 \le i \le t-1$,*

$$\alpha_i \ge \frac{t}{t-u}\left[(t-2u) - \frac{t(u+1)^2}{(t-u)^2-(t+1)-u(u+1)}\right] = a, \text{ say.} \quad (7.5.1)$$

□

From Lemma 7.5.2, the following result is immediate.

Theorem 7.5.1. *Suppose $d_{\text{plan}} \in \Omega_{t,\mu_1 t,t}$ is a balanced uniform design. Then a sufficient condition for $d_{\min}$ (and consequently, d_{imp}) to be connected is that*

$$(t-2u)\left[(t-u)^2-(t+1)-u(u+1)\right]-t(u+1)^2>0.$$

□

Example 7.5.2. Suppose $u=2$, that is, the dropouts occur in the last two periods. Then the condition of Theorem 7.5.1 simplifies to

$$t^3-9t^2+8t+12>0,$$

which holds for all $t\geq 8$. Therefore, if one starts with a design d_{plan} that is balanced and uniform with $t\geq 8$, then the design d_{min} remains connected even if all observations in the last two periods are lost. Similarly, if $u=1$, then the condition reduces to

$$t^3-5t^2+4>0.$$

This holds for all $t\geq 5$, implying that for all balanced uniform designs with $t\geq 5$, d_{min} remains connected if subjects drop out in the last period. □

When d_{min} is connected, Majumdar *et al.* (2008) used the following measure of the loss of precision in d_{imp} with respect to d_{plan} due to subject dropout:

$$L_{d_{\text{imp}}:d_{\text{plan}}}=1-\frac{\text{tr}(C^+_{d_{\text{plan}}})}{\text{tr}(C^+_{d_{\text{imp}}})},$$

where A^+ denotes the Moore-Penrose inverse of a matrix A (see Section 1.2). So by Lemma 7.5.1, the maximum loss of precision due to subject dropout for d_{plan}, denoted by $ML_{d_{\text{plan}}}$, is given by

$$ML_{d_{\text{plan}}}=L_{d_{\text{min}}:d_{\text{plan}}}=1-\frac{\text{tr}(C^+_{d_{\text{plan}}})}{\text{tr}(C^+_{d_{\text{min}}})}.$$

Clearly, $ML_{d_{\text{plan}}}\geq L_{d_{\text{imp}}:d_{\text{plan}}}$. An upper bound to $ML_{d_{\text{plan}}}$ was obtained by Majumdar *et al.* (2008) which is stated below.

Theorem 7.5.2. *Suppose $d_{\text{plan}}\in\Omega_{t,\mu_1 t,t}$ is a balanced uniform design. Then*

$$ML_{d_{\text{plan}}}\leq 1-\frac{a(t^2-t-1)}{t(t-2)(t+1)},\qquad(7.5.2)$$

where a is as in (7.5.1). □

It can be seen that for a given u, the bound in Theorem 7.5.2 decreases with t. For designs d_{plan} which are uniform, balanced and possess some additional nice combinatorial structures, Majumdar *et al.* (2008) gave sharper bounds than those in Theorem 7.5.2.

Anderson and Preece (2002) introduced a class of crossover designs called *locally balanced designs* and gave several methods of construction of these designs for $t(=p)$ an odd prime or prime power. Due to their nice combinatorial structure, these designs preserve some degree of "balance" even if certain latter periods are truncated. Anderson and Preece (2002) remarked that these designs had insurance against subjects dropping out, however, they did not study the statistical properties of these designs in the event of subject dropouts.

Bose and Bagchi (2008) studied the optimality aspects of the designs d_{min} when d_{plan} belongs to a certain class, say $\mathcal{D}_1$, of locally balanced designs, also known as transitive arrays (see Morgan and Uddin (1996), Majumdar and Martin (2004)). Anderson and Preece (2002) gave the construction method for these designs when $t(=p)$ is an odd prime and n is equal to $t(t-1)$. For details of the construction and examples, we refer to their paper. Bose and Bagchi (2008) showed that when d_{plan} is a member of $\mathcal{D}_1$, then d_{plan} is itself universally optimal for direct and carryover effects over the binary subclass in $\Omega_{t,n,t}$. Furthermore, d_{min} obtained from such a d_{plan} remains optimal over the binary subclass in $\Omega_{t,n,t-u}$ when d_{min} consists of $t-u$ periods with $t-u \geq 3$.

Bose and Bagchi (2008) also proposed some designs involving either $2t$ or $4t$ subjects, and showed that these designs are universally optimal for direct and carryover effects over the binary subclass in $\Omega_{t,n,t}$, and moreover, the corresponding d_{min} is optimal over a similar subclass when $t-u = (p+1)/2$. Thus though such a design, when used as d_{plan}, is less robust against dropouts than the ones in $\mathcal{D}_1$, it requires fewer subjects than a design in $\mathcal{D}_1$ for the same t.

For $p > 3$, we give below three examples of possible d_{plan} designs with $t = 5$ and $t = 7$. Here the designs d_{min} obtained from d_1 and d_2 are optimal if $t-u \geq 3$, while the d_{min} obtained from d_3 remains optimal if $t-u = 4$.

Example 7.5.3. $t = p = 5, n = 20$.

$$d_1 = \begin{array}{cccc} 1\,2\,3\,4\,5 & 2\,3\,4\,5\,1 & 1\,2\,3\,4\,5 & 2\,3\,4\,5\,1 \\ 2\,3\,4\,5\,1 & 4\,5\,1\,2\,3 & 5\,1\,2\,3\,4 & 5\,1\,2\,3\,4 \\ 5\,1\,2\,3\,4 & 5\,1\,2\,3\,4 & 2\,3\,4\,5\,1 & 4\,5\,1\,2\,3 \\ 3\,4\,5\,1\,2 & 1\,2\,3\,4\,5 & 4\,5\,1\,2\,3 & 3\,4\,5\,1\,2 \\ 4\,5\,1\,2\,3 & 3\,4\,5\,1\,2 & 3\,4\,5\,1\,2 & 1\,2\,3\,4\,5 \end{array}.$$

□

Example 7.5.4. $t = p = 7, n = 42.$

$$d_2 = \begin{array}{llllll} 1234567 & 1234567 & 1234567 & 1234567 & 1234567 & 1234567 \\ 2345671 & 7123456 & 3456712 & 6712345 & 4567123 & 5671234 \\ 7123456 & 2345671 & 6712345 & 3456712 & 5671234 & 4567123 \\ 3456712 & 6712345 & 5671234 & 4567123 & 7123456 & 2345671 \\ 6712345 & 3456712 & 4567123 & 5671234 & 2345671 & 7123456 \\ 4567123 & 5671234 & 7123456 & 2345671 & 3456712 & 6712345 \\ 5671234 & 4567123 & 2345671 & 7123456 & 6712345 & 3456712 \end{array}.$$

□

Example 7.5.5. $t = p = 7, n = 14.$

$$d_3 = \begin{array}{cccccccccccccc} 1&2&3&4&5&6&0&6&0&1&2&3&4&5 \\ 2&3&4&5&6&0&1&5&6&0&1&2&3&4 \\ 4&5&6&0&1&2&3&3&4&5&6&0&1&2 \\ 0&1&2&3&4&5&6&0&1&2&3&4&5&6 \\ 3&4&5&6&0&1&2&4&5&6&0&1&2&3 \\ 5&6&0&1&2&3&4&2&3&4&5&6&0&1 \\ 6&0&1&2&3&4&5&1&2&3&4&5&6&0 \end{array}.$$

□

The foregoing discussion in this section refers to the traditional model (1.3.3). Hedayat and Yan (2008) touched upon the problem of subject dropout under a model with self and mixed carryovers. They suggested the use of a d_{plan} given by an $OA_I(n, p, t, 2)$ with $p = 3$ or 4. When $p = 3$, this design is optimal if all p periods can be implemented (see Theorem 7.3.2). If the experiment can be continued only for $p'(< p)$ periods, then the resulting design is again an $OA_I(n, p', t, 2)$ and will be highly efficient.

7.6 Some Additional Comments

In this section, we give a brief account of some more developments that have a bearing on the subject matter of this book. Alongside the theoretical developments in the area of optimal crossover designs, several recent contributions center around developing suitable computer algorithms for the generation of optimal or efficient crossover designs and we begin with a description of these.

Eccleston and Street (1994) suggested a two-stage algorithm for generating efficient crossover designs. In the first stage, all or a large number of (M, S)-optimal row-column designs are generated. Note that a design is said to be (M, S)-optimal in a given class of competing designs if it has the minimum value of $\text{tr}(C^2)$ among those designs that have the maximum value of $\text{tr}(C)$, where C is the information matrix for a relevant set of parametric functions. The second stage consists of permuting the rows of each (M, S)-optimal design to find the design that minimizes a certain function depending on the number of times treatments preceded other treatments. The approach based on the (M, S)-optimality criterion is not very satisfactory as this criterion is really not an optimality criterion (see e.g., Cheng (1996)) and is rather a procedure for quickly identifying designs that might be optimal or highly efficient with respect to other meaningful criteria. The main advantage of the (M, S) criterion is that the computations involved are easy to carry out. However, as shown by Russell (1991), an optimal crossover design is not necessarily based on an (M, S)-optimal row-column design. See also Eccleston and Whitaker (1999) for one more algorithm.

Another algorithm proposed by Donev and Jones (1995) and Donev (1997) aims at iteratively improving a starting design by exchanging sequences of treatments with alternate sequences. The effect of each exchange is evaluated and the one that gives the maximum improvement in the value of a chosen optimality criterion is carried out. Several runs of the algorithm are made with different starting designs and the best design over all runs is finally recommended. The criterion chosen by these authors is the maximization of the average efficiency factor for direct effects. This method is computationally expensive and the use of this algorithm is limited by the fact that it becomes practically infeasible when a large number of candidate designs are to be evaluated.

John and Russell (2003) developed recursive formulae for computing the average efficiency factors for direct and carryover effect contrasts after a pair of treatments are interchanged in a starting design. These formulae are helpful for implementation in an interchange algorithm and result in quick generation of an optimal or near-optimal crossover design.

John, Russell and Whitaker (2004) provided a unified theory for the construction of crossover designs under a variety of models, all with uncorrelated errors and described an algorithm to obtain optimal or highly efficient designs under such models. This work was extended by Williams and John (2007), who provided an algorithm for generating efficient crossover designs when the errors either follow an autoregressive structure or a linear variance structure. For more details on these algorithms, the original sources may be consulted.

In the earlier chapters, some methods of construction of crossover designs have been described, which are optimal or efficient under different models. Several other methods of construction are available in the literature. In an early paper, Atkinson (1966) presented crossover designs with $p < t$. In Chapter 3, we have discussed a method of construction of Patterson designs through cyclic development of some initial treatment sequences. Designs constructed by a cyclic process were also proposed by Davis and Hall (1969) and Constantine and Hedayat (1982). Davis and Hall (1969) showed how cyclic incomplete block designs can be used to generate crossover designs. They observed that the ease in analysis and efficiency associated with cyclic incomplete block designs are retained when these are viewed as crossover designs and analyzed under a model suitable for crossover experiments. Some methods of construction of balanced crossover designs with $p < t$ were described by Hedayat and Afsarinejad (1975). Constantine and Hedayat (1982) reported families of crossover designs with $p < t$ which are balanced and can be constructed by a cyclic process. Balanced designs with $p < t$ can as well be obtained from the handcuffed designs of Lawless (1974). See also Lawless (1971), for a construction of balanced crossover designs via the use of BIB designs. The optimality aspects of these designs, however, have not yet been studied fully.

So far in this book, we have been concerned with crossover experiments in which it was implicitly assumed that the treatments do not have a factorial structure. Situations arise in practice where there is a need to investigate not only the individual but also the joint effects of two or more treatment factors. Mason and Hinkelmann (1971) were probably the first to consider the design and analysis of crossover experiments involving two or three factors at two levels each and two factors at three levels each. Fletcher and John (1985) introduced the notion of factorial structure in the context of crossover experiments and showed that a large class of crossover designs possess this desirable property. Subsequently, Fletcher (1987) used the generalized cyclic method of construction to obtain crossover designs for factorial experiments with $p < t$. As an example, consider the following design for a 2×3 factorial experiment, with $p = 4$ periods and $n = 12$:

$$\begin{array}{cccccccccccc}
00 & 01 & 02 & 10 & 11 & 12 & 00 & 01 & 02 & 10 & 11 & 12 \\
12 & 10 & 11 & 02 & 00 & 01 & 02 & 00 & 01 & 12 & 10 & 11 \\
01 & 02 & 00 & 11 & 12 & 10 & 11 & 12 & 10 & 01 & 02 & 00 \\
10 & 11 & 12 & 00 & 01 & 02 & 10 & 11 & 12 & 00 & 01 & 02
\end{array}.$$

Here, the first factor is at 2 levels, coded as 0 and 1 and the second factor is at 3 levels, 0, 1 and 2. All the designs considered by Fletcher (1987), including the one shown above have a cyclic structure, which in turn is helpful in the analysis and in the computation of efficiency factors. Lewis, Fletcher and Matthews (1988) obtained efficient crossover designs for experiments involving three or four periods and two treatment factors, where either both the factors have two levels or one has two levels and the other has three levels. More results on crossover designs with factorial treatments were obtained by Russell and Dean (1998). The optimality of these designs are yet to be fully explored, though most of the designs are reported to have high efficiency factors for the relevant parameters of interest.

Dean, Lewis and Chang (1999) described nested crossover designs suitable for experiments in which subjects are required to perform a series of tasks (say, b levels of a factor B) under a given set of experimental conditions in any one session. The different conditions are a levels of another factor, say A, and these conditions are changed from session to session. For instance, the different conditions could be different lighting conditions or different kinds of equipment. Since in such experiments, it is often difficult to change these conditions, each subject has to perform all the tasks under one set of conditions during a session. Thus factor B is nested within factor A. Dean *et al.* (1999) gave a series of nested crossover designs which are universally optimal for direct effects of the ab treatment combinations under a model that has only the first order carryover effects of factor B, apart from the other parameters as in model (1.2.1). They also present a subclass of designs which are additionally, universally optimal for the first order carryover effects of factor B.

Several tables of efficient or optimal crossover designs are available. An early informative table of such designs along with notes on steps of analysis and other information was prepared by Patterson and Lucas (1962). A large number of balanced and partially balanced designs, along with their respective efficiency factors were given by them. Some of these designs are now known to be optimal. Street, Eccleston and Wilson (1990) presented tables of small optimal or near-optimal crossover designs for $t = 2, 3$. These authors also listed designs with $t = 3 = p$, $4 \leq n \leq 10$, which maximize the average efficiency factor of the direct effects and presented a table of designs for some specific values of t, n and p which maximize the average efficiency factor of the carryover effects. Iqbal and Jones (1994) gave tables of efficient crossover designs with $t \leq 10$, $p \leq 8$, using sets of cyclic shifts. They also presented efficient strongly balanced designs and designs balanced for both first and second order carryover effects. Additionally, they gave efficient designs with two different period sizes. Tables of useful crossover designs also appear in Ratkowsky *et al.* (1992) and Jones and Kenward (2003).

We finally make some comments on the similarity of the crossover design model with some models that have been considered in other contexts. Models with carryover effects have also been used in the study of serially balanced sequences; see Bose (1996) and the references therein. Again, there are models that incorporate neighbor effects; for instance, there may be two effects carrying over from the neighboring plots (experimental units), usually from the plots on the right and left. Tools developed for such interference models are useful in the context of crossover designs, which have only one directional carryover effects. For more details on these models and related optimal designs, see Kunert and Martin (2000a).

References

Afsarinejad, K. (1983). Balanced repeated measurements designs. *Biometrika* **70**, 199–204.

Afsarinejad, K. (1985). Optimal repeated measurements designs. *Statistics* **16**, 563–568.

Afsarinejad, K. (1989). Circular balanced uniform repeated mesurements designs. *Statist. Probab. Lett.* **7**, 187–189.

Afsarinejad, K. (1990a). Circular balanced uniform repeated mesurements designs, II. *Statist. Probab. Lett.* **9**, 141–143.

Afsarinejad, K. (1990b). Repeated measurements designs–A review. *Comm. Statist. Theor. Meth.* **19**, 3985–4028.

Afsarinejad, K. and A. S. Hedayat (2002). Repeated measurements designs for a model with self and carryover effects. *J. Statist. Plann. Inference* **106**, 449–459.

Anderson, I. and D. A. Preece (2002). Locally balanced change-over designs. *Util. Math.* **62**, 33–59.

Archdeacon, D. S., J. H. Dinitz, D. R. Stinson and T. W. Tilson (1980). Some new row-complete Latin squares. *J. Combin. Theor.* **A29**, 395–398.

Armitage, P. and M. Hills (1982). The two-period crossover trial. *The Statistician* **31**, 119–131.

Atkinson, G. F. (1966). Designs for sequences of treatments with carryover effects. *Biometrics* **22**, 292–309.

Azzalini A. and A. Giovagnoli (1987). Some optimal designs for repeated measurements with autoregressive errors. *Biometrika* **74**, 725–734.

Bailey, R. A. and J. Kunert (2006). On optimal crossover designs when carryover effects are proportional to direct effects. *Biometrika* **93**, 613–625.

Balaam, L. (1968). A two-period design with t^2 experimental units. *Biometrics* **24**, 61–73.

Ballinger, C., R. M. Pickering, S. Bannister and D. L. McLellan (1995). Evaluating equipment for people with disabilities: user and technical perspective of commodes. *Clinical Rehab.* **9**, 157–166.

Barker, N., R. Hews, A. Huitson and J. Poloniecki (1982). The two period crossover trial. *Bull. Appl. Statist.* **9**, 67–116.

Bate, S. T. and B. Jones (2006). The construction of nearly balanced and nearly strongly balanced uniform cross-over designs. *J. Statist. Plann. Inference* **136**, 3248–3267.

Bate, S. T. and B. Jones (2008). A review of uniform cross-over designs. *J. Statist. Plann. Inference* **138**, 336–351.

Berenblut, I. I. (1964). Change-over designs with complete balance for first residual effects. *Biometrics* **20**, 707–712.

Berenblut, I. I. (1967). A changeover design for testing a treatment factor at four equally spaced levels. *J. Roy. Statist. Soc.* **B29**, 670–673.

Berenblut, I. I. (1968). Change over designs balanced for the linear component of first residual effects. *Biometrika* **55**, 297–303.

Berenblut, I. I. (1970). Treatment sequences balanced for the linear component of residual effects. *Biometrics* **26**, 154–156.

Berenblut, I. I. and G. I. Webb (1974). Experimental designs in the presence of autocorrelated errors. *Biometrika* **61**, 427–437.

Bishop, S. H. and B. Jones (1984). A review of higher order crossover designs. *J. Appl. Statist.* **11**, 29–50.

Bora, A. C. (1984). Change over designs with errors following a first order autoregressive process. *Austral. J. Statist.* **26**, 179–188.

Bora, A. C. (1985). Change over designs with first order residual effects and errors following a first order autoregressive process. *The Statistician* **34**, 161–173.

Bose, M. (1996). Some efficient incomplete block sequences. *Biometrika* **83**, 956–961.

Bose, M. and S. Bagchi (2008). Optimal crossover designs under premature stopping. *Util. Math.* **75**, 273–285.

Bose, M. and A. Dey (2003). Some small and efficient cross-over designs under a non-additive model. *Util. Math.* **63**, 173–182.

Bose, M. and A. Dey (2006). Combined intra-inter unit analysis of crossover designs and related optimality results. *J. Indian Soc. Agric. Statist.* **60**, 144–150.

Bose, M. and B. Mukherjee (2000). Cross-over designs in the presence of higher order carry-overs. *Austral. N. Z. J. Statist.* **42**, 235–244.

Bose, M. and B. Mukherjee (2003). Optimal crossover designs under a general model. *Statist. Probab. Lett.* **62**, 413–418.

Bose, M. and K. R. Shah (2005). Estimation of residual effects in repeated measurements designs. *Calcutta Statist. Assoc. Bull., Special Volume* **56**, 125–143.

Bose, M. and J. Stufken (2007). Optimal crossover designs when carry-over effects are proportional to direct effects. *J. Statist. Plann. Inference* **137**, 3291–3302.

Bose, R. C. and K. R. Nair (1941). On complete sets of Latin squares. *Sankhyā* **5**, 361–382.

Carriere, K. C. and G. C. Reinsel (1992). Investigation of dual-balanced crossover designs for two treatments. *Biometrics* **48**, 1157–1164.

Carriere, K. C. and G. C. Reinsel (1993). Optimal two-period repeated mesurements designs with two or more treatments. *Biometrika* **80**, 924–929.

Carriere, K. C. and R. Huang (2000). Crossover designs for two-treatment clinical trials. *J. Statist. Plann. Inference* **87**, 125–134.

Chassan, J. B. (1964). On the analysis of simple cross-overs with unequal numbers of replicates. *Biometrics* **20**, 206–208.

Cheng, C.-S. (1978). Optimality of certain asymmetrical experimental designs. *Ann. Statist.* **6**, 1239–1261.

Cheng, C.-S. (1996). Optimal design: Exact theory. In *Handbook of Statistics* **13** (C. R. Rao and S. Ghosh, Eds.). Amsterdam: North-Holland, pp. 977–1006.

Cheng, C.-S. and C. F. Wu (1980). Balanced repeated measurements designs. *Ann. Statist.* **8**, 1272–1283. Corrigendum: *ibid.* **11** (1983), 349.

Chow, S. C. and J. P. Liu (1992). On assessment of bioequivalence with high-order crossover designs. *J. Biopharmaceutical Statist.* **2**, 239–256.

Cochran, W. G. (1939). Long-term agricultural experiments. *J. Roy. Statist. Soc.* **6B**, 104–148.

Cochran, W. G., K. M. Autrey and C. Y. Cannon (1941). A double change-over design for dairy cattle feeding experiments. *J. Dairy Sc.* **24**, 937–951.

Cochran, W. G. and G. M. Cox (1957). *Experimental Designs*, 2nd ed. New York: Wiley.

Collombier, D. and I. Merchermek (1993). Optimal cross-over experimental designs. *Sankhyā* **B 55**, 249–261.

Constantine, G. and A. Hedayat (1982). A construction of repeated measurements designs with balance for residual effects. *J. Statist. Plann. Inference* **6**, 153–164.

Cotton, J. W. (1998). *Analyzing Within-Subjects Experiments.* Mahawah, New Jersey: Lawrence Erlbaum Assoc.

Cox, D. R. (1958). *Planning of Experiments.* New York: Wiley.

Cross, D. V. (1973). Sequential dependencies and regression in psychophysical judgements. *Percept. Psychophys.* **14**, 547–552.

Davis, A. W. and W. B. Hall (1969). Cyclic change-over designs. *Biometrika* **56**, 283–293.

Dean, A. M., S. M. Lewis and J. Y. Chang (1999). Nested changeover designs. *J. Statist. Plann. Inference* **77**, 337–351.

Dean, A. M. and D. Voss (1999). *Design and Analysis of Experiments.* New York: Springer.

Dénes, J. and A. D. Keedwell (1974). *Latin Squares and their Applications.* New York: Academic Press.

Dey, A. (1986). *Theory of Block Designs.* New York: Halsted.

Dey, A. and G. Balachandran (1976). A class of changeover designs balanced for first residual effects. *J. Indian Soc. Agric. Statist.* **28(2)**, 57–64.

Dey, A., V. K. Gupta and M. Singh (1983). Optimal change-over designs. *Sankhyā* **B45**, 233–239.

Dey, A. and R. Mukerjee (1999). *Fractional Factorial Plans.* New York: Wiley.

Donev, A. N. (1997). An algorithm for the construction of crossover trials. *Appl. Statist.* **46**, 288–298.

Donev, A. N. (1998). Crossover designs with correlated observations. *Stat. Med.* **8**, 249-262.

Donev, A. and B. Jones (1995). Construction of *A*-optimal cross-over designs. In *MODA4–Advances in Model Oriented Data Analysis* (C. P. Kitsos and W. G. Muller, Eds.). Heidelberg: Physica-Verlag, pp. 165–171.

Durier, C., H. Monod and A. Bruetschy (1997). Design and analysis of factorial sensory experiments with carryover effects. *Food Qual. Pref.* **8**, 131–139.

Ebbutt, A. F. (1984). Three period cross-over designs for two treatments. *Biometrics* **40**, 219–224.

Eccleston, J. A. and D. J. Street (1994). An algorithm for the construction of optimal or near-optimal change-over designs. *Austral. J. Statist.* **36**, 371–378.

Eccleston, J. A. and D. Whitaker (1999). On the design of optimal change-over experiments through multi-objective simulated annealing. *Statist. Comput.* **9**, 37–42.

Federer, W. T. (1955). *Experimental Design: Theory and Application.* New York: Macmillan.

Finney, D. J. (1956). Cross-over designs in bioassay. *Proc. Roy. Soc.* **B** **145**, 42–61.

Finney, D. J. (1978). *Statistical Method in Biological Assay*, 3rd ed. London: Griffin.

Fletcher, D. J. (1987). A new class of change-over designs for factorial experiments. *Biometrika* **74**, 649–654.

Fletcher, D. J. and J. A. John (1985). Changeover designs and factorial structure. *J. Roy. Statist. Soc.* **B** **47**, 117–124.

Freeman, G. H. (1959). The use of the same experimental material for more than one set of treatments. *Appl. Statist.* **8**, 13–20.

Gill, P. S. (1992). Balanced change-over designs for autocorrelated observations. *Austral. J. Statist.* **34**, 415–420.

Gill, P. S. and G. K. Shukla (1987). Optimal change-over designs for correlated observations. *Comm. Statist. Theor. Meth.* **16**, 2243–2261.

Godolphin, J. D. (2004). Simple pilot procedures for the avoidance of disconnected experimental designs. *Appl. Statist.* **53**, 133–147.

Grizzle, J. E. (1965). The two-period change-over design and its use in clinical trials. *Biometrics* **21**, 469–480. Corrigendum: *ibid.* **30** (1974), 727.

Gupta, S. and R. Mukerjee (1989). *A Calculus for Factorial Arrangements.* Lecture Notes in Statistics. New York: Springer-Verlag.

Hedayat, A. (1981). Repeated measurements designs, IV: Recent advances (with discussion). *Bull. Inter. Statist. Inst.* **XLIX**, 591–610.

Hedayat, A. and K. Afsarinejad (1975). Repeated measurements designs, I. In *A Survey of Statistical Designs and Linear Models* (J. N. Srivastava, Ed.). Amsterdam: North-Holland, pp. 229–242.

Hedayat, A. and K. Afsarinejad (1978). Repeated measurements designs, II. *Ann. Statist.* **6**, 619–628.

Hedayat, A. S., M. Jacroux and D. Majumdar (1988). Optimal designs for comparing test treatments with controls. *Statistical Sci.* **4**, 462–476.

Hedayat, A. S., N. J. A. Sloane and J. Stufken (1999). *Orthogonal Arrays: Theory and Applications.* New York: Springer.

Hedayat, A. S., J. Stufken and M. Yang (2006). Optimal and efficient crossover designs when subject effects are random. *J. Amer. Statist. Assoc.* **101**, 1031–1038.

Hedayat, A. S. and Z. Yan (2008). Crossover designs based on type I orthogonal arrays for a self and mixed carryover effects model with correlated errors. *J. Statist. Plann. Inference* **138**, 2201-2213.

Hedayat, A. S. and M. Yang (2003). Universal optimality of balanced uniform crossover designs. *Ann. Statist.* **31**, 978–983.

Hedayat, A. S. and M. Yang (2004). Universal optimality for selected crossover designs. *J. Amer. Statist. Assoc.* **99**, 461–466.

Hedayat, A. S. and M. Yang (2005). Optimal and efficient crossover designs for comparing test treatments with a control treatment. *Ann. Statist.* **33**, 915–943.

Hedayat, A. S. and M. Yang (2006). Efficient crossover designs for comparing test treatments with a control treatment. *Discrete Math.* **306**, 3112–3124.

Hedayat, A. and W. Zhao (1990). Optimal two-period repeated measurements designs. *Ann. Statist.* **18**, 1805–1816. Corrigendum: *ibid.* **20** (1992), 619.

Higham, J. (1998). Row-complete Latin squares of every composite order exist. *J. Combin. Des.* **6**, 63–77.

Hills, M. and P. Armitage (1979). The two-period crossover clinical trial. *Br. J. Clin. Pharmacol.* **8**, 7–20.

Hinkelmann, K. and O. Kempthorne (2005). *Design and Analysis of Experiments*, Vol. 2. New York: Wiley.

Iqbal, I. and B. Jones (1994). Efficient repeated measurement designs with equal and unequal period sizes. *J. Statist. Plann. Inference* **42**, 79–88.

Jacobson, N. (1964). *Lectures in Abstract Algebra*, Volumes 1–3. New York: Van Nostrand.

Jacroux, M. (1983). Some minimum variance block designs for estimating treatment effects. *J. Roy. Statist. Soc.* **B 45**, 70–76.

John, J. A. and M. H. Quenouille (1977). *Experiments: Design and Analysis*, 2nd ed. London: Griffin.

John, J. A. and K. G. Russell (2003). Optimising changeover designs using the average efficiency factors. *J. Statist. Plann. Inference* **113**, 259–268.

John, J. A., K. G. Russell and D. Whitaker (2004). CrossOver: An algorithm for the construction of efficient crossover designs. *Statist. Med.* **23**, 2645-2658.

Jones, B. and C. Deppe (2001). Recent developments in the design of cross-over trials: A brief review and bibliography. In *Recent Advances in Experimntal Designs and Related topics* (S. Altan and J. Singh, Eds.). New York: Nova Science Publishers, pp. 153–173.

Jones, B. and M. G. Kenward (2003). *Design and Analysis of Cross-over Trials*, 2nd ed. London: Chapman and Hall.

Jones, B., J. Kunert and H. P. Wynn (1992). Information matrices for mixed effects models with applications to the optimality of repeated measurements designs. *J. Statist. Plann. Inference* **33**, 261–274.

Jones, B., J. Wang, P. Jarvis and W. Byrom (1999). Design of cross-over trials for pharmacokinetic studies. *J. Statist. Plann. Inference* **78**, 307–316.

Kempton, R. A., S. J. Ferris and O. David (2001). Optimal change-over designs when carryover effects are proportional to direct effects of treatments. *Biometrika* **88**, 391–399.

Kenward, M. G. and B. Jones (1998). Crossover trials. In *Encyclopaedia of Statistical Sciences, Update Volume 2* (S. Kotz, C. B. Read and D. L. Banks, Eds.). New York: Wiley, pp. 167–175.

Kershner, R. P. and W. T. Federer (1981). Two-treatment crossover designs for estimating a variety of effects. *J. Amer. Statist. Assoc.* **76**, 612–619.

Kiefer, J. (1958). On the nonrandomized optimality and randomized nonoptimality of symmetric designs. *Ann. Math. Statist.* **29**, 675–699.

Kiefer, J. (1975). Construction and optimality of generalized Youden designs. In *A Survey of Statistical Designs and Linear Models* (J. N. Srivastava, Ed.). Amsterdam: North-Holland, pp. 333–353.

Koch, G. G., I. A. Amara, B. W. Brown, T. Colton and D. B. Gillings (1989). A two-period crossover design for the comparison of two active treatments and placebo. *Statist. Med.* **8**, 487–504.

Kunert, J. (1983). Optimal design and refinement of the linear model with applications to repeated measurements designs. *Ann. Statist.* **11**, 247–257.

Kunert, J. (1984a). Designs balanced for circular carry-over effects. *Comm. Statist. Theor. Meth.* **13**, 2665–2671.

Kunert, J. (1984b). Optimality of balanced uniform repeated measurements designs. *Ann. Statist.* **12**, 1006–1017.

Kunert, J. (1985). Optimal repeated measurements designs for correlated observations and analysis by weighted least squares. *Biometrika* **72**, 375–389.

Kunert, J. (1991). Cross-over designs for two treatments and correlated errors. *Biometrika* **78**, 315–324.

Kunert, J. (1998). Sensory experiments as crossover studies. *Food Qual. Pref.* **9**, 243–253.

Kunert, J. and R. J. Martin (2000a). On the determination of optimal designs for an interference model. *Ann. Statist.* **28**, 1728–1742.

Kunert, J. and R. J. Martin (2000b). Optimality of type I orthogonal arrays for cross-over models with correlated errors. *J. Statist. Plann. Inference* **87**, 119–124.

Kunert, J. and J. Stufken (2002). Optimal crossover designs in a model with self and mixed carryover effects. *J. Amer. Statist. Assoc.* **97**, 898–906.

Kurkjian, B. and M. Zelen (1962). A calculus for factorial arrangements. *Ann. Math. Statist.* **33**, 600–619.

Kurkjian, B. and M. Zelen (1963). Applications of the calculus for factorial arrangements, I: Block and direct product designs. *Biometrika* **50**, 63–73.

Kushner, H. B. (1997a). Optimality and efficiency of two-treatment repeated measurements designs. *Biometrika* **84**, 455–468. Corrigendum: *ibid.* **86** (1999), 234.

Kushner, H. B. (1997b). Optimal repeated measurements designs: the linear optimality equations. *Ann. Statist.* **25**, 2328–2344. Corrigendum: *ibid.* **27** (1999), 2081.

Kushner, H. B. (1998). Optimal and efficient repeated measurements designs for uncorrelated observations. *J. Amer. Statist. Assoc.* **93**, 1176–1187.

Kushner, H. B. (1999). *H*-symmetric optimal repeated measurements designs. *J. Statist. Plann. Inference* **76**, 235-261.

Lakatos, E. and D. Raghavarao (1987). Undiminished residual effects designs and their suggested applications. *Comm. Statist. Theor. Meth.* **16**, 1345–1359.

Laska, E. and M. Meisner (1985). A variational approach to optimal two-treatment crossover designs: application to carryover-effect models. *J. Amer. Statist. Assoc.* **80**, 704–710.

Laska, E., M. Meisner and H. B. Kushner (1983). Optimal crossover designs in the presence of carryover effects. *Biometrics* **39**, 1087–1091.

Lawless, J. F. (1971). A note on certain types of BIBDs balanced for residual effects. *Ann. Math. Statist.* **42**, 1439–1441.

Lawless, J. F. (1974). On the construction of handcuffed designs. *J. Combin. Theor.* **A 16**, 87–96.

Laywine, C. F. and G. L. Mullen (1998). *Discrete Mathematics Using Latin Squares.* New York: Wiley.

Lewis, S. M., D. J. Fletcher and J. N. S. Matthews (1988). Factorial cross-over designs in clinical trials. In *Optimal Design and Analysis of Experiments* (Y. Dodge, V. V. Fedorov and H. P. Wynn, Eds.). Amsterdam: North-Holland, pp. 133–140.

Low, J. L., S. M. Lewis and P. Prescott (1999). Assessing robustness of crossover designs to subjects dropping out. *Statist. Comput.* **9**, 219–227.

Low, J. L., S. M. Lewis and P. Prescott (2003). An application of Pólya theory to cross-over designs with dropout. *Util. Math.* **63**, 129–142.

Low, J. L., S. M. Lewis, B. D. McKay and P. Prescott (1994). Computational issues for cross-over designs subject to dropout. *COMPSTAT* (Vienna, 1994). Heidelberg: Physica, pp. 423–428.

Lucas, H. L. (1957). Extra-period Latin-square change-over designs. *J. Dairy Sc.* **40**, 225–239.

Magda, C. G. (1980). Circular balanced repeated measurements designs. *Comm. Statist. Theor. Meth.* **9**, 1901–1918.

Majumdar, D. (1988). Optimal repeated measurements designs for comparing test treatments with a control. *Comm. Statist. Theor. Meth.* **17**, 3687–3703.

Majumdar, D. (1996). Optimal and efficient treatment-control designs. In *Handbook of Statistics* **13** (S. Ghosh and C. R. Rao, eds.), Amsterdam: North-Holland, pp. 1007–1053.

Majumdar, D., A. M. Dean and S. M. Lewis (2008). Uniformly balanced repeated measurements designs in the presence of subject dropout. *Statist. Sinica* **18**, 235–253.

Majumdar, D. and R. J. Martin (2004). Efficient designs based on orthogonal arrays of type I and II for experiments using units ordered over time or space. *Statist. Methodology* **1**, 19–35.

Markiewicz, A. (1997). Properties of information matrices for linear models and universal optimality of experimental designs. *J. Statist. Plann. Inference* **59**, 127–137.

Martin, R. J. and J. A. Eccleston (1998). Variance-balanced change-over designs for dependent observations. *Biometrika* **85**, 883–892.

Mason, J. M. and K. Hinkelmann (1971). Change-over designs for testing different treatment factors at several levels. *Biometrics* **27**, 430–435.

Matthews, J. N. S. (1987). Optimal crossover designs for the comparison of two treatments in the presence of carryover effects and autocorrelated errors. *Biometrika* **74**, 311–320. Corrigendum: *ibid.* **75** (1988), 396.

Matthews, J. N. S. (1988). Recent developments in crossover designs. *Internat. Statist. Rev.* **56**, 117–127.

Matthews, J. N. S. (1990). Optimal dual-balanced two treatment crossover designs. *Sankhyā* **B 52**, 332–337.

Matthews, J. N. S. (1994a). Multi-period crossover designs. *Statistical Methods in Medical Research* **3**, 383–405.

Matthews, J. N. S. (1994b). Modelling and optimality in the design of crossover studies for medical applications. *J. Statist. Plann. Inference* **42**, 89–108.

McNutty, P. A. (1986). Spatial Velocity Induction and Reference Mark Density. Unpublished Ph.D. thesis, University of California, Santa Barbara.

Mendelsohn, N. S. (1968). Hamilton decomposition of the complete directed n-graph. In *Theory of Graphs* (P. Erdös and G. Katona, Eds.). Amsterdam: North-Holland, pp. 237–241.

Mielke, P. W., Jr. (1974). Squared rank test appropriate to weather modification cross-over design. *Technometrics* **16**, 13–16.

Morgan, J. P. and N. Uddin (1996). Optimal blocked main effects plans with nested rows and columns and related designs. *Ann. Statist.* **24**, 1185–1208.

Mukerjee, R. (1979). Inter-effect orthogonality in factorial experiments. *Calcutta Statist. Assoc. Bull.* **28**, 83–108.

Mukerjee, R. (1980). Further results on the analysis of factorial experiments. *Calcutta Statist. Assoc. Bull.* **29**, 1–26.

Mukhopadhyay, A. C. and R. Saha (1983). Repeated measurements designs. *Calcutta Statist. Assoc. Bull.* **32**, 153–168.

Namboodiri, N. K. (1972). Experimental designs in which each subject is used repeatedly. *Psy. Bull.* **77**, 54–64.

Patterson, H. D. (1951). Change-over trials (with discussion). *J. Roy. Statist. Soc.* **B 13**, 256–271.

Patterson, H. D. (1952). The construction of balanced designs for experiments involving sequences of treatments. *Biometrika* **39**, 32–48.

Patterson, H. D. (1970). Non-additivity in change-over designs for a quantitaive factor at four levels. *Biometrika* **57**, 537–549.

Patterson, H. D. (1973). Quenouille's changeover designs. *Biometrika* **60**, 33–45.

Patterson, H. D. and H. L. Lucas (1959). Extra-period change-over designs. *Biometrics* **15**, 116–132.

Patterson, H. D. and H. L. Lucas (1962). *Change-over Designs.* North Carolina Agricutural Experiment Station Tech. Bull. No. 147.

Pigeon, J. G. and D. Raghavarao (1987). Crossover designs for comparing treatments with a control. *Biometrika* **74**, 321–328.

Pocock, S. (1983). *Clinical Trials.* New York: Wiley.

Prescott, P. (1999). Construction of uniform-balanced cross-over designs for any odd number of treatments. *Statist. Med.* **18**, 265–272.

Pukelsheim, F. (1993). *Optimal Design of Experiments.* New York: Wiley.

Quenouille, M. H. (1953). *The Design and Analysis of Experiment.* London: Griffin.

Raghavarao, D. (1971). *Constructions and Combinatorial Problems in Design of Experiments.* New York: Wiley.

Raghavarao, D. (1990). Crossover designs in industry. In *Statistical Design and Analysis of Industrial Experiments* (S. Ghosh, Ed.). New York: Marcel Dekker, pp. 517–530.

Rao, C. R. (1961). Combinatorial arrangements analogous to orthogonal arrays. *Sankhyā* **A23**, 283–286.

Ratkowsky, D. A., M. A. Evans and J. R. Alldredge (1992). *Cross-over Experiments: Design, Analysis, and Application.* New York: Marcel Dekker.

Rupp, M. E., T. Fitzgerald, S. Puumala, J. R. Anderson, R. Craig, P. C. Iwen, D. Jourdan, J. Keuchel, N. Marion, D. Peterson, L. Sholtz and V. Smith (2008). Prospective, controlled, cross-over trial of alcohol-based hand gel in critical care units. *Infect. Control Hosp. Epidemiol.* **29**, 8–15.

Russell, K. G. (1991). The construction of good change-over designs when there are fewer units than treatments. *Biometrika* **78**, 305–313.

Russell, K. G. and A. M. Dean (1998). Factorial cross-over designs with few subjects. *J. Combin. Inform. System Sci.* **23**, 209–233.

Schifferstein, H. N. V. and I. M. Oudejans (1996). Determination of cumulative successive contrasts in saltiness intensity judgements. *Percept. Psychophys.* **58**, 713–724.

Sen, M.[1] and R. Mukerjee (1987). Optimal repeated measurements designs under interaction. *J. Statist. Plann. Inference* **17**, 81–91.

Senn, S. (1992). Is the simple carryover model useful? *Statist. Med.* **11**, 715–726.

Senn, S. J. (1994). The *AB/BA* crossover: past, present and future? *Statistical Methods in Medical Research* **3**, 303–324.

Senn, S. J. (1997). Cross-over trials. In *Encyclopaedia of Biostatistics* **2** (P. Armitage and T. Colton, Eds.). New York: Wiley, pp. 1033–1049.

Senn, S. J. (2000). Crossover design. In *Encyclopaedia of Biopharmaceutical Statistics* (S. C. Chow, Ed.). New York: Marcel Dekker, pp. 142–149.

Senn, S. (2003). *Cross-over Trials in Clinical Research*, 2nd ed. Chichester, England: Wiley.

Senn, S. J. and H. Hildebrand (1991). Crossover trials, degrees of freedom, the carryover problem and its dual. *Statist. Med.* **10**, 1361–1374.

Shah, K. R., M. Bose and D. Raghavarao (2005). Universal optimality of Patterson's designs. *Ann. Statist.* **33**, 2854–2872.

Shah, K. R. and B. K. Sinha (1989). *Theory of Optimal Designs.* Lecture Notes in Statistics. New York: Springer.

Shah, K. R. and B. K. Sinha (2002). Universal optimality for the joint estimation of parameters. Unpublished manuscript.

Sheehe, P. R. and D. J. Bross (1961). Latin squares to balance immediate residual, and other order, effects. *Biometrics* **17**, 405–414.

[1]Mausumi Sen is the maiden name of Mausumi Bose

Stevens, W. L. (1939). The completely orthogonalized Latin square. *Ann. Eugen.* **9**, 82.

Street, D. J., J. A. Eccleston and W. H. Wilson (1990). Tables of small optimal repeated measurements designs. *Austral. J. Statist.* **32**, 345–359.

Stufken, J. (1991). Some families of optimal and efficient repeated measurements designs. *J. Statist. Plann. Inference* **27**, 75–83.

Stufken, J. (1996). Optimal crossover designs. In *Handbook of Statistics* **13** (S. Ghosh and C. R. Rao, Eds.). Amsterdam: North-Holland, pp. 63–90.

Takeuchi, K. (1961). On the optimality of certain type of PBIB designs. *Rep. Statist. Appl. Res. Un. Japan Sci. Engg.* **8**, 140-145.

Ting, C. P. (2002). Optimal and efficient repeated measurements designs for comparing test treatments with a control. *Metrika* **56**, 229–238.

Wakeling, I. N. and H. J. H. MacFie (1995). Designing consumer trials balanced for first and higher order carry-over effects when only a subset of k samples from t may be tested. *Food Qual. Pref.* **6**, 299–308.

Wang, L. L. (1973). A test for the sequences of a class of finite graphs with two generators. *Notices Amer. Math. Soc.* **20**, 73T–A275.

Willan, A. R. and J. L. Pater (1986). Carryover and the two-period crossover clinical trial. *Biometrics* **42**, 593–599.

Willey, R. F., I. W. B. Grant and S. J. Pocock (1976). Comparison of cardiorespiratory effects of oral salbutamol and pirbuterol in patients with bronchial asthma. *Br. J. Clin. Pharmacol.* **3**, 595–600.

Williams, E. J. (1949). Experimental designs balanced for the estimation of residual effects of treatments. *Austral. J. Sci. Res. A* **2**, 149–168.

Williams, E. J. (1950). Experimental designs balanced for pairs of residual effects. *Austral. J. Sci. Res. A* **3**, 351–363.

Williams, E. R. and J. A. John (2007). Construction of crossover designs with correlated errors. *Austral. N. Z. J. Statist.* **49**, 61–68.

Wu, C. F. (1980). On some ordering properties of the generalized inverse of nonnegative definite matrices. *Linear Alg. Appl.* **32**, 49–60.

Yang, M. and M. Park (2007). Efficient crossover designs for comparing test treatments with a control treatment when $p = 3$. *J. Statist. Plann. Inference* **137**, 2056–2067.

Yang, M. and J. Stufken (2008). Optimal and efficient crossover designs for comparing test treatments to a control treatment under various models. *J. Statist. Plann. Inference* **138**, 278–285.

Zelen, M. and W. T. Federer (1964). Applications of the calculus for factorial arrangements, II. Designs for two-way elimination of heterogeneity. *Ann. Math. Statist.* **35**, 658–672.

Index